# PROTOPLASMATOLOGIA
## HANDBUCH
## DER PROTOPLASMAFORSCHUNG

HERAUSGEGEBEN VON

### L. V. HEILBRUNN und F. WEBER
PHILADELPHIA  GRAZ

MITHERAUSGEBER

W. H. ARISZ - GRONINGEN · H. BAUER - WILHELMSHAVEN · J. BRACHET-
BRUXELLES · H. G. CALLAN - ST. ANDREWS · R. COLLANDER - HELSINKI ·
K. DAN - TOKYO · E. FAURÉ - FREMIET - PARIS · A. FREY - WYSSLING - ZÜRICH ·
L. GEITLER - WIEN · K. HÖFLER - WIEN · M. H. JACOBS - PHILADELPHIA ·
D. MAZIA - BERKELEY · A. MONROY - PALERMO · J. RUNNSTRÖM - STOCKHOLM ·
W. J. SCHMIDT - GIESSEN · S. STRUGGER - MÜNSTER

### BAND II
### CYTOPLASMA

E

### CYTOPLASMA-OBERFLÄCHE

4

THE ENZYMOLOGY OF THE CELL SURFACE

5

TENSION AT THE CELL SURFACE

### WIEN
### SPRINGER-VERLAG
1954

# THE ENZYMOLOGY OF THE CELL SURFACE

BY

## ASER ROTHSTEIN

ROCHESTER, NEW YORK

WITH 21 FIGURES

# TENSION AT THE CELL SURFACE

BY

## E. NEWTON HARVEY

PRINCETON, NEW JERSEY,

WITH 13 FIGURES

WIEN

SPRINGER-VERLAG

1954

ISBN-13: 978-3-211-80345-5     e-ISBN-13: 978-3-7091-5449-6
DOI: 10.1007/978-3-7091-5449-6

# The Enzymology of the Cell Surface

By

ASER ROTHSTEIN

Division of Pharmacology, Department of Radiation Biology
University of Rochester School of Medicine and Dentistry
Rochester, New York

With 21 Figures

## Contents

# Introduction

In recent years there has been a convergence of interest on the part of the biochemists on one hand and the biologists and cytologists on the other. The biochemist has become increasingly aware that biochemical activities of the cell are intimately related to the structure of the cell. The biologist has become increasingly aware that the morphological elements of the cell possess important metabolic functions. The common interest of the biochemist and the biologist has resulted in the present popularity of research concerned with the localization of enzyme systems within the cell. Considerable impetus has also been given to such studies by the development of techniques which allow the isolation of cellular elements such as the mitochondria, microsomes, chloroplasts and nuclei. In addition new cytological techniques such as electron microscopy, autoradiography, and ultra-violet microscopy, as well as the more recent modifications of staining procedures, have increased our knowledge of the metabolic geography of the cell. Not only has it become apparent that specific enzymes are intimately associated with certain structural elements of the cell, but it is evident that the architecture of the cell plays an important role in the regulation of metabolic activities.

The present monograph is concerned with the enzymology of the surface of the cell. Until recently, for a number of reasons, little consideration has been given to the possibility that the cell surface might participate directly in enzymic reactions. In the first place, there have been few correlations between any cellular structure and any metabolic activity. Secondly, few obvious techniques were available which could successfully demonstrate the existence of enzymes in the cell surface. Prior to 1940, there were a mere handful of papers dealing with the subject. Thirdly, most of the studies of the cell surface have been concerned with its more obvious properties, its structure, composition, electrical properties, mechanical properties, and most particularly, its permeability properties. None of these properties are obviously associated with surface enzymes.

What are the functions of the cell surface that may require the presence of enzymes? From a purely hypothetical point of view, the cellular activities in which cell surface enzymes are potentially functional can be divided into two categories; first, those activities related to the production and maintenance of the cell surface structure itself, and second, those activities which are concerned with interactions between the cell and its environment. During growth of a cell, as it enlarges, new structural material must be synthesized and assembled in the cell membrane. Although it is conceivable that new structural material is synthesized in the interior of the cell, it is difficult to visualize its complete synthesis and final assembly in the periphery of the cell, without the assistance of enzymes at the site. In the case of non-growing cells, although no net synthesis of new structural material is necessary, the turnover of the various components by a continuous degradation and resynthesis must be considered as a possibility. Interactions between the cell and its environment, from the present point of view, are of three types: (1) the digestion or breakdown of extracellular substances, (2) the synthesis of extracellular polymerized substances such as proteins, fibers, cell walls and slimes, and (3) the active transfer of substances from the environment to the interior of the cell or from the interior of the cell to the environment.

Until recently there were few unequivocal demonstrations of cell surface enzymes. However, there were a number of suggestions that enzymatic activity of the cell surface might offer a logical explanation for certain phenomena. For example, QUASTEL (1926) and QUASTEL and WOOLDRIDGE (1927 a and 1927 b) studied the reduction of certain dyes by intact cells of microorganisms. They concluded that the dehydrogenases concerned must be located on the outer surface of the cell. CLARK (1937) suggested that metabolic activities occur on the cell surface and that these can be modified by drugs, thus explaining the sequence of events in the action of certain drugs. Cell surface enzymes have also been proposed to explain the mechanisms of sugar resorption (HÖBER 1946; ROSENBERG and WILBRANDT 1952; WILBRANDT 1954). In recent years, the number of hypotheses predicated on the existence of cell surface enzymes has increased markedly, particularly in relation to active transport or secretion phenomena which will be discussed later in more detail. The hypothetical existence of surface enzymes, in the absence of direct experimental verification, is by itself not completely convincing no matter how reasonable the hypothesis may be. However, in the recent literature, it has been clearly demonstrated in a number of cases that specific enzymes are located on the cell surface. In addition it has been shown beyond question that certain metabolic activities are associated with the cell surface, even though the specific enzymes involved are not known. The unequivocal demonstration of the existence of certain enzymes on the cell surface serves

1*

to strengthen but not to prove the various hypotheses predicated on surface enzyme activity.

In summary the material to be discussed in this monograph includes demonstrations of specific enzymes, enzymes hypothesized to satisfy the needs of a theory, and phenomena which probably involve surface enzymes but which have not as yet been critically investigated. It is beyond the scope of this article to deal with all of the many interactions between the cell and its environment in which surface enzymes are potentially involved. The greatest emphasis will be laid on those cases in which cell surface enzymes have been definitely implicated, with secondary emphasis on those cases in which the data are at least suggestive of surface enzyme activity.

The fact that there are more theories about cell surface enzymes than there are clear-cut experimental demonstrations, not only verifies the old observation that it is easier to make suggestions than it is to do experiments, but also speaks for the technical difficulties involved. In fact, technical difficulties rather than disinterest have held up the fuller development of research in the field of cell surface enzymology. Because methodology has been a severe handicap, those techniques that have proven useful will be discussed in order to demonstrate the experimental bases on which the concept of cell surface enzymes depends.

## Methods of Localizing Enzymes on the Surface of the Cell

The study of the enzymology of a number of sub-cellular structures has progressed rapidly because methods have been developed for their isolation, in quantity, without obvious damage to structure or enzymic activity. Unfortunately, no one has yet invented adequate methods of "skinning" cells, or of isolating intact cell surfaces. To be sure, some work has been done with the stroma of the red cell and with the giant nerve fiber of the squid with the axoplasm extruded. However, these preparations have only a limited usefulness. Cytological techniques, including auto-radiography have been useful in some cases, but they too have many limitations. Most of the data on the cell surface have been derived from "in vivo" studies of the cell membrane using an intact living cell. Here the problem is to differentiate the reactions occurring at the surface of the cell from those occurring in the interior of the cell. A number of procedures have been useful. In some cases it has been possible to show that either the substrate or the products of the reaction have not penetrated into the interior of the cell. In other cases it is possible to demonstrate that certain reactions are influenced by substances (inhibitors or accelerating substances) in the extracellular environment. The enzyme is thereby localized in the surface of the cell where it is accessible to the extracellular factors. Finally, enzymes have been "placed" on the cell surface on the basis of indirect evidence or on the basis of reasonable assumptions, without direct proof.

## Isolation of the Cell Surface

Only in a few cases has it been feasible to physically isolate the cell surface from the rest of the cytoplasm. The red blood cell can be hemolyzed and washed, leaving a residue, the ghost, or stroma, which presumably contains the surface structure of the cell. However, ghosts may contain up to $1/4$ of the original volume of the cell, depending on the method of preparation. With prolonged washing the volume may be reduced to $1/20$th of that of the original cell, corresponding to a layer of 250 Å thick (PONDER 1948; PARPART and BALLENTINE 1952). In the case of the giant axon of the squid, it is possible to squeeze out the axoplasm, the fluid content of the interior of the nerve, and thereby make a crude separation (BOELL and NACHMANSOHN 1940). Isolation of the cell surface has also been reported in protozoa (SEAMAN 1951) and in bacteria (MITCHELL and MOYLE 1951). It is obvious in each case that a complete separation of an intact surface from the cytoplasm is difficult. Nevertheless, any separation of cell surface material, no matter how crude, may lead to useful information. In the case of the red cell stroma, a rather constant composition can be obtained by adequate washing. However, much of the material originally present in the surface may have been washed out. For this reason it should be borne in mind that a positive finding of enzyme activity in the isolated structure is meaningful but a negative finding might be due to the loss of activity during the isolation procedure.

Another procedure which may turn out to be useful is based only partly on a physical separation of the cell surface structure. Certain reactions in sugar uptake have been characterized as occurring at the surface of the living yeast cell by techniques which will be described. A cell-free insoluble preparation of yeast has been made by drying, lyophilizing and extraction procedures (ROTHSTEIN, DEMIS, and BRUCE 1954). This preparation contains a structural entity, with a volume of less than 10% of that of the original cell, and possessing many of the same biochemical and permeability properties toward sugars attributed to the surface of the living cell. Thus, even though the structure has never been completely separated from the other cell residues, it is believed to be part of the cell surface structure. The evidence will be discussed later.

## Cytological Procedures

In recent years a number of cytological procedures have been developed for localizing certain enzyme activities within the cell. In regard to the cell surface, two staining procedures which have been particularly useful are the phosphatase method of GOMORI (1939) and choline esterase method of KOELLE and FRIEDENWALD (1949). Cytological procedures must always be interpreted with caution because of the possibility of artifacts, a problem which is reviewed elsewhere (DANIELLI 1953). Furthermore, attempts to localize enzymes at the cell surface by cytological techniques are limited since the cell membrane is beyond the resolving power of the microscope.

Nevertheless, the appearance of the typical enzyme staining in a band around the periphery of the cell indicates that the enzyme is located either in the cell surface or at least in the underlying cortical region of the cell.

Recently, reactions of phosphate uptake by muscle have been localized in the cell surface by cytological techniques involving radio-autography with $P^{32}$ (CAUSEY and HARRIS 1951). The muscle cell is sufficiently large so that there is no doubt from the autographs that there is a peripheral fixation of the phosphate. Again, the resolution is inadequate to indicate the specific peripheral structures involved.

## Non-Penetrating Substrates or Products

Any substrate which cannot pass through the cell membrane into the cytoplasm, but which can be chemically changed in the presence of cells, must be altered either by a secreted enzyme or by an enzyme bound on the surface of the cell. The possibility that a secreted enzyme is responsible can be eliminated if no enzymatic activity can be found in the medium immediately after the cells are removed; that is, if the cells must be present for the reaction to proceed. An example of the action of cells on a non-penetrating substrate is the splitting of phosphate compounds by yeast cells (ROTHSTEIN and MEIER 1948). Yeast cells are impermeable to organic phosphates, yet can hydrolyse them, liberating orthophosphate and an organic moiety, both of which can be recovered in the medium. Furthermore, if an organic phosphate compound labeled with $P^{32}$ is used, then the compound is split with no mixing of the liberated labeled orthophosphate with the orthophosphate already present in the cytoplasm. Thus the substrate has had no contact with the general cytoplasm of the cell, yet the hydrolysis only occurs in the presence of cells.

## Non-Penetrating Inhibitors or Stimulating Agents

In a number of cases it has been possible to find inhibitors or accelerating agents which influence the biochemical activities of cells even though the agents can be shown not to penetrate into the cell. The absence of penetration has been shown in a number of ways. For example, uranyl ion is bound by the cell and inhibits glucose uptake by yeast. Yet uranium inhibition is reversed by the addition of a low concentration of inorganic phosphate to the medium, despite the fact that much higher concentrations of orthophosphate are already present in the interior of the cell. The uranium must therefore be bound to the cell surface where it is exposed to the extracellular rather than intracellular phosphate (ROTHSTEIN and LARRABEE 1948). In another case, spores of fungus are exposed to 0.1 N acid. The invertase activity of the spores is destroyed, yet the normal germination and glucose utilization is unimpaired. Thus the acid does not penetrate into the cytoplasm itself. The invertase must be peripherally located, where it is exposed to the external environment (MANDELS 1953 a).

If a non-penetrating agent influences a cellular activity, then it is evident that the process which is inhibited or accelerated must occur at the cell membrane. It is still necessary to show that this process is enzymic in nature, rather than an effect on permeability or on some other non-enzymic transport mechanism. Sometimes it is possible to establish identity by comparing the action of the agent on the intact cell and on a purified enzyme. WILKES and PALMER (1932) showed that the pH effect on the invertase activity of the living yeast cell was identical with that on the isolated enzyme. It would be coincidence indeed if some permeability factor possessed exactly the same pH-activity curve as does invertase.

If the effects of non-penetrating agents on cells cannot be equated with effects on specific enzymes, then at least they can be shown to be similar to the effects of these agents on enzymes in general. Such parallelism does not necessarily *prove* that an enzyme is involved but it is supportive evidence.

## Kinetic Analyses

The dynamics of a reaction at the cell surface can help in identifying the reaction. If the surface reaction simply involves the penetration of the membrane by diffusion, then the rate of uptake of a substance should be proportional to the concentration gradient. If, however, a chemical combination with a surface component is essential to the reaction, then as the concentration of the reacting substance is increased, the rate will reach a constant level, associated with saturation of the surface component. Such a saturation effect is observed in sugar uptake by red blood cells (WIL-BRANDT 1954; LeFEVRE 1954) and by yeast (HURWITZ and ROTHSTEIN 1951). The kinetic approach alone will not differentiate between a surface enzyme and a non-enzymic surface carrier substance which combines with the substrate and carries it across the membrane.

## Temperature Effects

Temperature dependence is of limited usefulness in determining the nature of cell surface reactions. Enzyme reactions have a high temperature coefficient whereas diffusion possesses a low one. However, DANIELLI (1943) has shown that the diffusion of polar substances through a lipid membrane may be highly temperature dependent.

## Specificity

Substrate specificity is a property of enzymes. If a cell surface reaction possesses the same specificity as a known enzyme, then it seems likely that that enzyme is involved in the reaction. For example, yeast is impermeable to galactose, sorbose and arabinose (CONWAY and DOWNEY 1950 a; ROTHSTEIN and MEIER, Unpublished Observations), but takes up glucose, fructose and mannose.  Yeast hexokinase possesses the same specificity (Kunitz and McDonald 1946; Slein, Cori and Cori 1950), suggesting that this enzyme is involved in sugar uptake.

A non-enzymic surface carrier substance might also possess substrate specificity.  Thus the interpretation of data on substrate specificity must be made with caution.

## Precursor Relationships with Isotopes

In isotope experiments certain criteria have been set up which serve to establish with some certainty the precursor relationships between various compounds (Zilversmit, Entenman, and Fishler 1943).  The criteria have to do with the sequence of the labeling in the compounds.  In simplest terms, those compounds that show the highest degree of labeling (relative specific activity) soonest are probably the precursors of the other compounds.  In human red blood cells (Gourley 1952 a; Prankerd and Altman 1954) and in sea-urchin eggs (Lindberg 1950), it has been found in studies with $P^{32}$ labeled orthophosphate that organic phosphate compounds such as ATP can qualify as precursors for the cellular inorganic phosphate.  The data indicate that the extracellular phosphate does not come directly in contact with the cellular orthophosphate, but is first incorporated into ATP.  Presumably the sequence of reactions is:

$$PO_4 \text{ of the medium} \rightarrow ATP \rightarrow PO_4 \text{ of the cell.}$$

The phosphorylation reactions which esterify the extracellular phosphate to ATP must take place in the periphery of the cell rather than the interior, else the orthophosphate from the medium could equilibrate directly with that inside the cell without ATP acting as an intermediate.

There are some precautions that must be observed in interpreting precursor relationships in cellular compounds.  In the first place, a given compound in the cell may be compartmentalized.  It may be actively turning over in one location but not in another.  On isolation of the compound by chemical means the active fraction is mixed with the inactive fraction, with consequent dilution of the isotope used to measure the turnover.  In mammalian tissues another complication exists.  The cells are in equilibrium with the substances in the interstitial spaces but direct measurement cannot be made of the interstitial fluid.  Therefore, calculations must be made to correct for the substances in the interstitial space, or attempts must be made to wash out substances in the interstitial space.  Either procedure is subject to certain limitations (Ennor and Rosenberg 1954 a and 1954 b).

## Concentration Gradients

During the accumulation of substances by the cell, the inward flow may occur in response to the concentration gradient.  The flow can be maintained if the cell continuously converts the substance inside the cell to another compound, thereby maintaining the concentration gradient. Such a process has been termed "trapping" (ROSENBERG and WILBRANDT 1952). In other cases, the substances are moved *against* the concentration gradient, a process known as "active transport". In such cases a "pumping system" must be present which is energized by the metabolism of the cell. Although the exact nature of the pump is not known in any particular case, the proposed mechanisms usually involve a coupling of enzymes to the pumping systems in the cell surface. Thus a clear cut demonstration of active transport raises the possibility that surface enzymes may be involved.

# Definition of "Cell Surface"

The term "cell surface" has often been used interchangeably with another term, the plasma membrane. A general discussion of the properties of the plasma membrane is given by DANIELLI (1951). It is a thin membrane, perhaps a 100 Å units thick, composed of lipids, proteins, nucleic acids and perhaps carbohydrates. It serves as the permeability barrier of the cell. Underneath the plasma membrane there is a gelatinous layer of cytoplasm, the cortex of the cell. In some cells, the plasma membrane is the outermost part of the cell, in others there may be a layer of material outside of the plasma membrane, such as the cell walls of plant cells and microorganisms, the capsules of bacteria, and the jelly coats of eggs.

Unfortunately the methods for localizing enzymes in the cell surface do not always allow a rigorous cytological definition of the "cell surface". The enzymes might actually be in the cell wall structure, the plasma membrane, or the cortex of the cell. For example, the limiting factor in cytological technique is the resolution of the microscope. At best it can be stated on the basis of these methods, that the enzymes are located in the periphery of the cell. The methods based on non-penetrating substrates indicate that the reactions are proceeding in a very small fraction of the cytoplasm at the periphery of the cell. For example, when ATP labeled with $P^{32}$ is split by cells, there is no significant dilution of the liberated phosphate by cellular phosphate. Thus the splitting must occur on the outside of the cell. However, a peripheral zone of the cell, constituting 2 to 3% of the cytoplasm could be involved and the equilibration of this zone would not be detected. In other studies, it is shown that certain reactions are influenced by the extracellular rather than intracellular pH. There is, however, no precise method of determining just where the barrier is located, outside of which the extracellular pH is the predominant factor. In other words, the term "cell surface" as used in this monograph has no precise cytological meaning, nor is it a surface in a geometrical sense. It is rather a peripheral layer, or zone of the cell. The extent to which this zone can be restricted to a particular cytological structure like the plasma membrane depends entirely on the method used and its precision or resolution.

# Surface Enzymes which Chemically Alter Substances in the Medium

The type of surface enzyme which is easiest to demonstrate is one which acts on an extracellular substrate. In the simplest case, neither the substrate nor its products can penetrate the cell. Thus at any particular time both the unused substrate and the products can be quantitatively recovered and identified in the medium. In other cases the substrate cannot penetrate the cell nor can it be utilized by the cell, but one or all of the products can be taken up and utilized. The enzymes involved can be classed as surface-bound "digestive" enzymes with the function of converting non-utilizable substances into utilizable products.

Because surface enzymes are exposed to the extracellular rather than intracellular environment, they can often be characterized in some detail without removal from the living cell. For example the pH response of the enzyme can be determined in the living cell by simply altering the extracellular pH. The property of the enzyme so determined does not differ appreciably from that found with a purified enzyme.

## Polysaccharide Splitting Enzymes

There are many observations in the older literature which show that living yeast cells possess considerable invertase and maltase activity, that is, ability to split sucrose and maltose into hexoses. The invertase is firmly bound to the cellular structure and is not liberated into the medium by living yeast cells. Even in disrupted cells, the enzyme is bound to insoluble cellular debris and it is only released and solubilized on autolysis or after digestion with papain (Thorsell and Myrback 1951). The first observations which threw some light on the location of the enzymes in the cell were those of Willstätter and Lowry (1925). They found that yeast cells treated for about an hour with 0.2 to 0.3 N $H_2SO_4$ lost their invertase and maltase activity, but that the ability of the cells to ferment glucose, or to grow and divide was not impaired. This observation has been confirmed (unpublished data). Mandels (1953 a) has recently carried out similar studies with spores of *Myrothecium verrucaria* and also of *A. luchuensis.* Treatment with 0.1 HCl destroys the invertase activity of both kinds of spores, and the cellobiose, trehalase and maltase of the *A. luchuensis,* but does not influence the viability or metabolism. If the cytoplasm were accessible to 0.1 N HCl it does not seem likely that the cells would remain viable, nor would they be capable of metabolizing glucose. The susceptibility of the fermentative enzymes to acid is demonstrated by the fact that soluble zymase preparations of yeast capable of fermenting glucose are irreversibly inactivated by acid treatment. Thus, in the living cell, the cytoplasm is protected from the extracellular acid by a peripheral barrier. The carbohydrases, invertase, maltase, cellobiose and raffinase are not protected. They must be located outside of the permeability barrier, directly exposed to the extracellular environment.

A somewhat different approach was used by Wilkes and Palmer (1932). They found that the invertase activity of the living yeast cell was remarkably dependent on the pH of the medium. In fact, the pH activity curve for the living yeast cell was almost identical with that for a purified invertase preparation from the same yeast cells (Fig. 1). Assuming that the internal pH must be at least somewhat independent of the environmental pH, Wilkes and Palmer suggested that the invertase must be located in the periphery of the cell where it would be exposed to the extracellular pH rather than the intracellular pH. Their basic assumption that the internal pH of the yeast cell is at least somewhat independent of the external pH over the range studied, although not verified experimentally at the time, was essentially correct. It has since been shown that exposure of yeast to a pH range of 1 to 10 does not alter the average internal pH by more than a few tenths of a pH unit (Rothstein and Demis 1954). It could be argued that the extracellular pH alters the permeability of the cell membrane to sucrose. However, it is unlikely that the pH curve for permeability could be identical with that for purified invertase. Furthermore, evidence will be presented later which indicates that sucrose does not penetrate into the yeast cell (Demis, Rothstein, and Meier 1954). Therefore, permeability cannot be a factor.

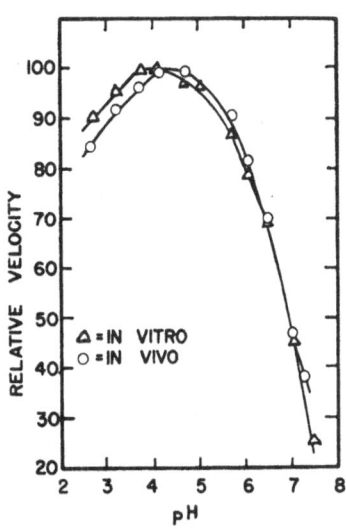

Fig. 1. The effect of pH on invertase activity of yeast cells and of isolated enzyme. (Wilkes and Palmer 1932; courtesy J. gen. Physiol.)

Myrback and Oertenblad (1936 and 1937) and Myrback and Vasseur (1943) have studied the lactase and trehalase activities of living yeast cells. They found that trehalose added to a suspension of yeast is hydrolyzed, yet the cell itself normally contains relatively high concentrations of trehalose in its cytoplasm. Therefore, it was suggested that the enzyme and the substrate were separated by the cell membrane. Furthermore, the fermentation of trehalose was found to possess a narrow pH optimum at about pH 5.0 (Myrback and Vasseur 1943), about the same as that for the activity of isolated trehalase. In contrast the fermentation of glucose was found to be relatively independent of pH over the range of 2–3 to beyond 7. Thus the fermentation of trehalose is not direct but must be preceded by its hydrolysis by a cell surface enzyme. In the case of lactase activity of the cell, the effects of pH were similar to those found by Wilkes and Palmer for invertase. Myrback and Vasseur concluded using the same rationale that this enzyme must also be located on the cell surface.

More recently, Mandels (1951) has studied the effect of pH on the invertase activities of spores of the fungus *Myrothecium verrucaria*. As in the case of yeast, the pH curve for the invertase activity of the intact spores is the same as that for the isolated enzymes, with an optimum at pH 3.5–4.0. Again, assuming that the pH within the spores is relatively constant and not subject to variations which occur in the environment, MANDELS suggested that the enzyme must be located on the cell surface.

Further evidence for the surface location of invertase in yeast is presented by DEMIS, ROTHSTEIN and MEIER (1954). They point out that sucrose can be split almost 300 times as rapidly as glucose can be taken up and fermented, and that the products of hydrolysis, glucose and fructose, can be quantitatively recovered in the medium. If the invertase were located inside the cell, sucrose would have to diffuse into the cell and the products, glucose and fructose, would have to diffuse out almost 300 times as rapidly as the known maximum rates of glucose and fructose uptake. Furthermore, by volume of distribution studies it was shown that during hydrolysis of sucrose by yeast cells, neither the sucrose nor products of its hydrolysis appear in the cellular water to any measureable extent. One further criterion of cell surface location of the enzyme was applied. The invertase activity of the cell could be completely inhibited by $1 \times 10^{-4}$ M uranyl nitrate. However, the inhibition could be reversed by the addition of orthophosphate. For example, $3 \times 10^{-5}$ uranyl nitrate inhibited to the extent of 37%. The addition of $1.2 \times 10^{-4}$ M orthophosphate reduced the inhibition to 9%. In view of the fact that the orthophosphate content of the yeast cytoplasm is between 1 and $2 \times 10^{-2}$ M in this yeast, it must be concluded that both the uranium and the enzyme that it inhibits, are exposed to the external rather than to the internal phosphate concentrations.

Certain yeasts which lack invertase activity will, if exposed to the purified enzyme, adsorb it firmly so that it cannot be washed out, but they only do so during active fermentation of glucose (OPARIN and YURKEVICH 1949). Yeasts which already have a high invertase activity do not adsorb invertase to such a great extent. Thus it is possible for cells to "acquire" surface bound enzymes. The yeast populations used were non-growing and no data are reported concerning the "acquired" invertase activity after growth and cell division. Invertase activity has also been observed at the cell surface in plant cells (BURSTROM 1941; DORMER and STREET 1949).

What is the function of the carbohydrases of the cell surface? DEMIS, ROTHSTEIN, and MEIER (1954) have pointed out that sucrose utilization by the cells is a two-step process, first a splitting to glucose and fructose, followed by metabolism of these products. Direct utilization of sucrose, which occurs in other cells (DOUDOROFF 1951), can only play an insignificant role in baker's yeast. Invertase in this case is a surface-bound, digestive enzyme. The same is probably true of the other carbohydrases. For example, MYRBACK and OERTENBLAD (1937) have demonstrated that hydrolysis of trehalose by surface trehalase precedes the fermentation of this sugar.

In addition to the enzymes which hydrolyze di- and trisaccharides, cells possess enzymes which can split the various high molecular weight polymers of sugar. However, there have been few investigations concerned with the cytological location of these enzymes. VASSEUR (1951) has suggested that the jelly splitting enzymes of sea-urchin sperm are located on the cell surface. Other sperm possess depolymerizing enzymes, but these may be secreted enzymes. A number of bacteria can depolymerize hyaluronic acid. In most cases the hyaluronidase is found in the culture medium, whether because of secretion or lysis of some of the cells is unknown. However, in young cultures of certain organisms the enzyme is bound to the cells (MEYER 1954). In view of the very high molecular weight of hyaluronic acid (as high as 1,000,000), it seems unlikely that this substance could penetrate into the cells. It seems probable therefore that hyaluronidase is located on the cell surface.

## Phosphatases

The staining technique of GOMORI (1939), and its modifications has been applied to many cells and tissues. A definite localization of alkaline phosphatase has been found at the periphery as well as in the interior of the cells of a wide variety of tissues. A number of investigators, for example, have demonstrated heavy phosphatase staining at the brush borders on the lumenal side of the cells of the intestinal and renal epithelia (GOMORI 1939; BOURNE 1943 and 1944; DEMPSEY and DEANE 1946; EMMEL 1946; Ross and ELY 1949). Other observers have noted a peripheral localization of alkaline phosphatase in cells of such diverse structures as the nephredial systems of invertebrates, silk glands of spiders and caterpillars, salivary glands of *Drosophila* (BRADFIELD 1949 and 1950), as well as in isolated cells such ast yeast (NICKERSON, KRUGELIS, and ANDRESEN 1948) and red blood cells (CLARKSON and MAIZELS 1952).

DANIELLI (1952) and BRADFIELD (1949 and 1950) point out that those cells to which an active secretory function can be attributed seem to have especially large amounts of peripherally located alkaline phosphatases. They believe that this correlation is more than a coincidence and that the surface phosphatases play an essential role in the secretory mechanism. ROSENBERG and WILBRANDT (1952) also discuss the possibility that the surface phosphatases are concerned in the active transport of glucose into the cell, a concept that will be discussed at greater length in a following section.

Surface phosphatases of yeast have been investigated in some detail by ROTHSTEIN and MEIER (1948 and 1949). They found that organic phosphates such as ATP, when added to a suspension of living yeast cells, were hydrolyzed with the appearance of the products in the medium. For each molecule of ATP added, one molecule of adenylic acid and 2 of orthophosphate were recoverable in the medium. Proof that the phosphatases were located in the surface of the cell was obtained by the use of ATP and glucose phosphates labeled with $P^{32}$. Fig. 2 shows such an experiment with ATP. The splitting to adenylic acid and orthophosphate is shown by the disappearance, within an hour, of almost all of the labile phosphate (7 minute hydrolyzable in $1 N$ HCl at $100^{\circ}$ C.) and the equivalent appearance of orthophosphate. However, no change occurs in ester phosphate, pentose or nitrogen, indicating that the adenylic acid moiety remains intact. The $P^{32}$ activity remains entirely in the medium, with none detectable in the cells. If the splitting of ATP took place in the interior of the cell, then some of the labeled orthophosphate liberated from ATP would mix with the cellular orthophosphate. In fact, ten times as much orthophosphate is hydrolyzed from the added ATP as is already present in the cells, yet no labeling of the cell phosphate occurs. Thus the phosphatases are not in the interior of the cell, nor are they excreted into the medium as shown by the absence of activity in the medium. They are bound on the surface of the cell.

Experiments were carried out with sugar phosphates labeled with $P^{32}$ with similar results. No mixing occurred between the phosphate released by hydrolysis of the substrate and that already present in the cell.

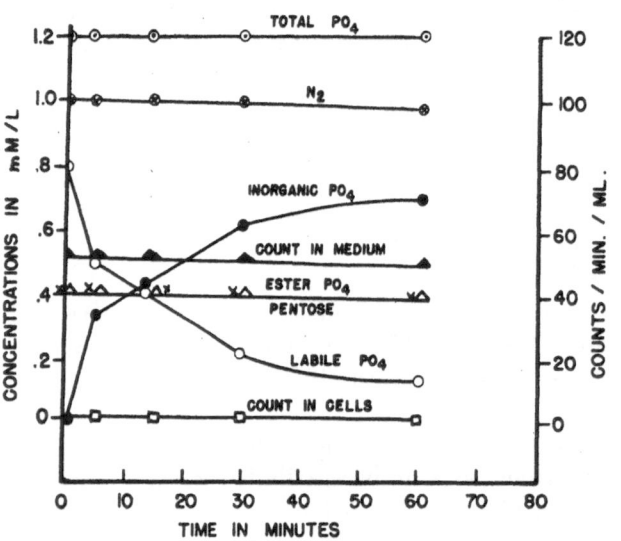

Fig. 2.  Changes in various constituents in the medium during hydrolysis of ATP containing $P^{32}$.

The yeast concentration was 5 mg./ml. and the pH was 3.5.
(Rothstein and Meier 1948; courtesy J. cellul. a. comp. Physiol.)

However, with the sugar phosphates, the sugar moiety is taken up by the cells and fermented. Only the phosphate is recovered in the medium.

Because the phosphatases are located on the surface of the cell, exposed to the external environment, their properties can be determined "in vivo". Among the substrates which are hydrolyzed in addition to ATP, glucose-1-phosphate and glucose-6-phosphate, are hexose diphosphate, ADP, phenyl phosphate, glycerophosphate, inorganic tripolyphosphate, pyrophosphate, and metaphosphate. Of the substances tested, only adenylic acid was not split. The pH curves for various substrates have optima in the region of pH 3.0 to 4.0. The curves are rather narrow for most substrates, but are somewhat broader for the sugar phosphates. There is little activity for any substrate at pH's higher than 7.0. The splitting of ATP fits the Michaelis-Menten kinetics. The Km is $1.6 \times 10^{-4}$ and Vm 0.52 micromols/hour/mg. of yeast, wet weight. Although the enzyme is inhibited by the products of the reaction, it is not destroyed during the experiment.

From competition experiments with pairs of substrates, from the pH curves, and from inhibition studies with molybdate, the conclusion can be drawn that no single phosphatase is responsible for the hydrolysis of all substrates tested. Rather, there is a family of enzymes with a fair degree of substrate specificity.

The cell-surface phosphatases of yeast were found to play a purely digestive function. None of the organic phosphates tested could penetrate into the living yeast cell, nor could they be utilized metabolically, unless they were first hydrolyzed by the phosphatases. For example, with sugar phosphate as the substrate, the rate of fermentation, although considerably slower than that with free glucose as a substrate, is exactly equal to the rate of dephosphorylation by phosphatase. Furthermore, the effect of external pH on phosphatase activity and on the fermentation of sugar phosphate is exactly the same (Fig. 3). It is apparent that the surface phosphatases can convert a substance which the cell cannot utilize (glucose phosphate) into substances which the cell can utilize (free glucose and orthophosphate). Although phosphate was not taken up in the experiments cited above, it can be actively absorbed by the cell under the proper conditions (see detailed discussion on phosphate uptake).

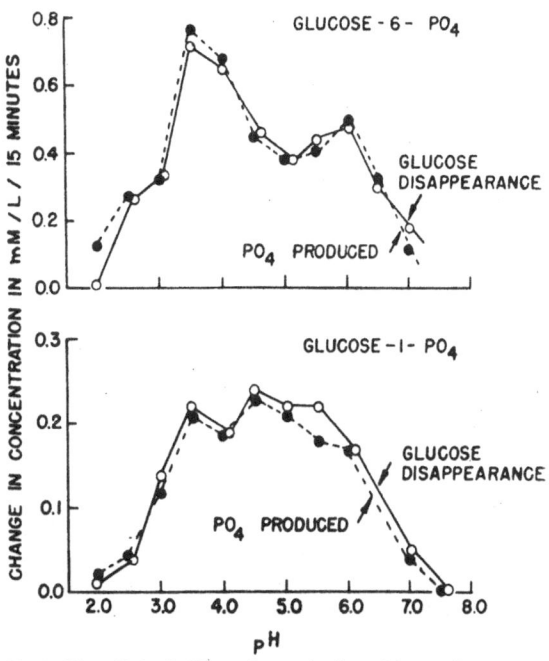

Fig. 3. The effect of pH on the production of inorganic phosphate and on the changes in total sugar when sugar phosphates are added to a yeast suspension.

The yeast concentration was 40 mg./ml. and the substrate concentration $3 \times 10^{-3}$ M.

(ROTHSTEIN and MEIER 1949; courtesy J. cellul. a. comp. Physiol.)

It has already been mentioned that phosphatases have been implicated as factors in active secretory functions of the cell (DANIELLI 1952). The acid-phosphatases of the yeast cells surface seem to have no such functions as revealed by studies with the phosphatase inhibitors, molybdate and tungstate. The surface-phosphatases are remarkably sensitive to molybdate ion and to tungstate. Concentrations of molybdate as low as $1 \times 10^{-7}$ M will block glycerophosphate hydrolysis to the extent of 40%. Although the hydrolysis of each compound requires a different molybdate concentration for inhibition, the hydrolysis of all the compounds tested is completely blocked by $1 \times 10^{-4}$ M molybdate (Fig. 4).

Using concentrations of molybdate that completely block the action of the surface phosphatases, little or no effect was observed on the ability of the cells to take up and ferment or respire glucose or to synthesize or degrade glycogen. Nor was there any effect on the ability of the cells to actively take up phosphate or to take up potassium in exchange for hydrogen ion. In brief, the phosphatases seem to play no role in carbohydrate metabolism nor in the associated transport of ions such as potassium, hydrogen and phosphate. The only measureable effect associated with the inhibition of the phosphatases was the inability of the cells to utilize sugar phosphates.

It should be mentioned that Sperber (1942), studying the uptake of thiamine pyrophosphate (TPP) by living yeast cells suggested that a surface phosphatase was responsible for the splitting of the compound to free thiamine plus orthophosphate and that the same enzyme, by a trans-phosphorylation reaction, was responsible for the uptake of thiamine and its resynthesis by the cell to the pyrophosphate. Although Westenbrink, Steyn-Parve, and Veldman (1947) have thrown considerable doubt on the concept of TPP synthesis by the phosphatase, Sperber's suggestion that TPP phosphatase is located on the cell surface is certainly justified in the light of later work.

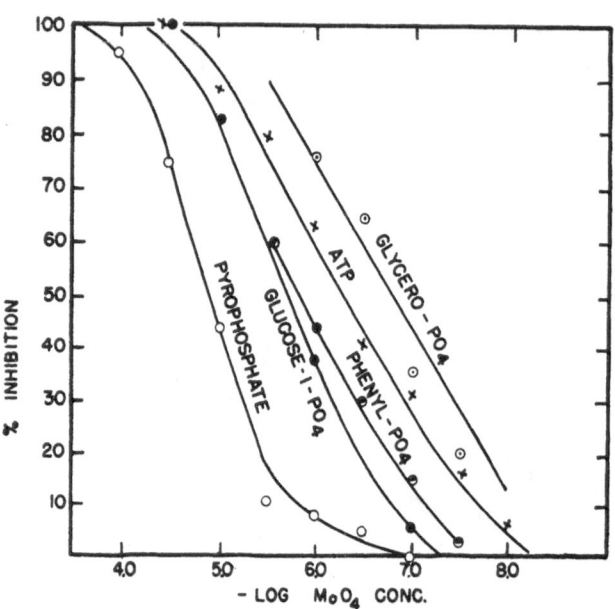

Fig. 4. Inhibition by molybdate of the hydrolysis of various phosphate compounds by yeast cells.

The inhibition was calculated from the inorganic phosphate liberated in 10 minutes by a yeast suspension containing 10 mg. wet weight per milliliter of suspension. Substrate concentrations were $10^{-3}$ M; the pH 3.5

(Rothstein and Meier 1949; courtesy J. cellul. a. comp. Physiol.)

Surface phosphatase of yeast have also been studied by Derrick, Miller, and Sevag (1953) using immunological techniques. Rabbit antiserum prepared against both purified yeast phosphatase, and against intact yeast cells will completely inhibit the activity of phosphatase in pure preparations or in the living yeast cell. In view of the very high molecular weight of the antibodies it seems unlikely that they could penetrate into the interior of the cell. Thus they must be inhibiting an enzyme on the outside surface of the cell. The inhibition of the yeast phosphatase by antibodies is a competitive one, reversible by high concentrations of the substrate.

Phosphatase on the surface of the red blood cell have been investigated by Clarkson and Maizels (1952) and by Herbert (1952). Their results are in essential agreement. Clarkson and Maizels found that intact red cells hydrolyze ATP with the consequent appearance in the medium of two equivalents of orthophosphate. Based on the observation of Hevesy (1947) that red cells are impermeable to phosphate esters, including ATP, it is concluded that the ATP splitting enzyme is located on the surface of the cell. This conclusion was confirmed by observations made on the red cell-stroma and on the hemolysates of the cells after removal of stroma. The ATP-ase activity is confined to the stroma, with none in the hemolysate.

The stroma also contains activity toward diphosphoglycerate if a co-factor in cell hemolysates is added. Activity toward diphosphoglycerate, inorganic pyrophosphate and triphosphate was found in the hemolysate but not in the stroma.

HERBERT confirmed most of the findings of CLARKSON and MAIZELS. He found that the stroma actually had about twice the ATP-ase activity of the intact cells suggesting that some of the stroma enzyme was not exposed to the environment in the intact cell. The ATP-ase of both stroma and the intact cell was very sensitive to copper. However, the addition of hemolysate of red cells reversed the copper inhibition indicating that in the living cell the copper was acting at the cell surface, out of contact with the cytoplasmic contents. Stroma was found to split ADP and pyrophosphate, but not fructose monophosphate, or phosphoglycerate. It is of some interest that the Michaelis constant for the stroma phosphatase is very similar to that found for yeast surface phosphatase, $2.4 \times 10^{-4}$ compared to $1.6 \times 10^{-4}$. However, the pH optima are quite different 7.0 as compared to 3.5.

The function of the surface phosphatases of the red cell is not specifically known.

Phosphatases have been localized in the surface of intestinal cells of rats by studies using $P^{32}$ labeled phosphate esters (ROTHSTEIN, MEIER, and SCHARFF 1953). The experiments were done with both "in vivo" and "in vitro" intestinal loops of starved (48 hour) male rats. Sugar phosphates as well as other phosphate compounds are rapidly hydrolyzed in the

Table 1. *Hydrolysis of Glucose-1-Phosphate in Intestinal Loops.*

| Loop No. | Phosphates | | | Glucose | | | Recovery of Total P | Recovery of Total Glucose |
|---|---|---|---|---|---|---|---|---|
| | Free | Bound | Total | Free | Bound | Total | | |
| | $\mu$M | $\mu$M | $\mu$M | $\mu$M | $\mu$M | $\mu$M | % | % |
| 1 . . . . . . . . | 78 | 10 | 88 | 33 | 10 | 43 | 92 | 45 |
| 2 . . . . . . . . | 49 | 47 | 96 | 31 | 48 | 79 | 101 | 83 |
| 3 . . . . . . . . | 37 | 57 | 94 | 36 | 58 | 94 | 98 | 98 |
| 4 . . . . . . . . | 50 | 42 | 92 | 29 | 42 | 71 | 96 | 74 |
| Av. . . . . . . . | 53 | 39 | 92 | 32 | 40 | 72 | 96 | 75 |
| Injected . . . | 0 | 95 | 95 | 0 | 95 | 95 | | |

(ROTHSTEIN, MEIER and SCHARFF 1953; courtesy Amer. J. Physiol.)

intestinal loops with the consequent appearance of free glucose and free phosphate in the lumen. Some of the glucose is subsequently absorbed, but all but 5% of the orthophosphate is recoverable in the lumen (Table 1). The rate of hydrolysis of the phosphate compound is about the same in all parts of the intestine and is not dependent on a maintained blood flow.

The uptake of glucose, however, shows a gradient along the length of the intestine, being highest at the duodenal end. It is very dependent on the adequate blood flow. In experiments with $P^{32}$-labeled glucose-1-phosphate (Table 2), there was no dilution of the labeled phosphate in the lumen

Table 2. *Recovery of Phosphate from $P^{32}$ Labeled Glucose-1-Phosphate Placed in Intestinal Loops.*

| Loop No. | Phosphate Analyses | | | $P^{32}$ Activity | | | |
| | Free | Bound | Total | Loop Contents | | Ashed Intestine Count | Total Recovery Count |
| | | | | Count | Specific Activity | | |
| | $\mu$M | $\mu$M | $\mu$M | c./m. | | c./m. | % |
| 1........... | 51 | 45 | 96 | 1342 | .99 | 57 | 99 |
| 2........... | 76 | 18 | 94 | 1283 | .96 | 84 | 96 |
| 3........... | 73 | 24 | 97 | 1324 | .97 | 114 | 102 |
| 4.......... | 67 | 24.8 | 91 | 1270 | .99 | 150 | 100 |
| Av. ........ | 66.8 | 26.8 | 94.5 | 1304 | .985 | 101 | 99,3 |
| Injected .... | 0 | 100 | 100 | 1416 | 1.00 | | |

Data from 4 rats are averaged.

(ROTHSTEIN, MEIER and SCHARFF 1953; courtesy Amer. J. Physiol.)

during hydrolysis of the compound, despite the fact that a large amount of phosphate is always present in the intestinal cells. If the hydrolysis occurred inside the cells, followed by a movement of some of the liberated orthophosphate back into the lumen, some mixing of cellular orthophosphate with $P^{32}$-labeled orthophosphate (derived from glucose-1-phosphate) would result, thereby decreasing its specific activity. It was therefore concluded that the splitting occured either at the cell surface or in the lumen. Because no enzyme activity was ever found in the lumen contents, it was concluded that the phosphatase is bound on the cell surface.

The surface phosphatases of the intestine appear to be digestive enzymes. Organic posphates such as sugar phosphates are not directly absorbed to an apreciable degree, but once hydrolyzed the sugar moiety is absorbed. The rate of splitting of sugar phosphate by the surface phosphatases is more rapid than the rate of absorption of sugar. In consequence free glucose accumulates in the lumen to some extent. This is the reason that the rate of sugar absorption from sugar phosphate is just as rapid as from free sugar as found by MATHIEU (1935).

The giant nerve fiber of the squid has been studied for ATP-ase activity by LIBET (1948). Using the technique of BOELL, and NACHMANSOHN (1940) for separating the axoplasm from the sheath, he found almost 99% of the enzyme activity of the nerve was located in the sheath. No specific function has been attributed to the ATP-ase of the nerve sheath.

## Choline Esterase

Enzymes capable of splitting acetyl choline are found widely distributed in tissues and in body fluids. There are at least two kinds of choline esterase. One found exclusively in red blood cells and in conducting tissue is relatively specific for the choline esters. It has been called "true choline esterase". The esterases of other tissues and of blood serum are relatively non-specific, hydrolyzing other esters as well (NACHMANSOHN and WILSON 1951). In regard to cell surface, it is the former enzyme which is of interest, for BOELL and NACHMANSOHN (1940) have shown in the giant axon of the squid that it is distributed almost entirely in the sheath of the nerve, with no appreciable activity in the axoplasm. In contrast, the respiratory enzymes and succinic dehydrogenase are found exclusively in the axoplasm (NACHMANSOHN and STEINBACH 1942; NACHMANSOHN et al. 1943). Similarly, the choline esterase activity of intact red blood cells has been found to be associated almost entirely with the stroma of the red cell with no loss of activity when the cell is hemolyzed and the hemolysate washed away (BRAUER and ROOT 1945; PALEUS 1947). MARNAY and NACHMANSOHN (1938) studied the choline esterase activity of different parts of frog muscle and found a correlation between the extent of the activity and the number of motor end plates in the different parts. They suggested, therefore, that the enzyme was localized at the motor end plate. Similar studies were carried out with the gastrocnemius of guinea pig with the same result (COUTEAUX and NACHMANSOHN 1938). Recently a staining technique has been developed for choline esterase (KOELLE and FRIEDENWALD 1949) which has confirmed the localization of this enzyme at the end plate. Using the same staining technique, SAWYER, DAVENPORT, and ALEXANDER (1950) and DENZ (1953) confirmed this result and also found a high concentration of the enzyme at the surface of the ganglion cell.

It should be kept in mind that the motor end plate, although it is a peripherally located struture, is not strictly a surface structure. It is a complex, containing nuclei and perhaps mitochondria.

It is of interest that the ciliate protozoan, *Tetrahymena gelii* S. contains a specific choline esterase. SEAMAN (1951), using homogenization and centrifugation procedures, has been able to separate the pellicle or outer structure from the remainder of the cells. He found almost all of the choline esterase activity in the surface structure.

The specific function of the cell surface choline esterase of conducting tissues is intimately connected with the function of its substrate acetyl choline. It is beyond the scope of the present paper to discuss in any detail the more recent concepts concerning the role acetyl choline in nerve physiology. Detailed information can be found elsewhere (NACHMANSOHN and WILSON 1951). It is sufficient to point out that acetyl choline seems to be definitely implicated in the transmission of impulses across the synapses of the autonomic ganglia, and as well, in transmission at the myoneural junction. However, transmission at the various central synapses may not involve acetyl choline. The choline esterase of the ganglionic synapses and of the myoneural junctions has the function of destroying acetyl choline. The activity of the enzyme therefore determines the duration of the acetyl choline effect. In normal function it seems likely that the choline esterase activity is sufficiently high to destroy the acetyl choline within the refractory period of the nerve fiber.

In the opinion of Nachmansohn and Wilson (1951), the surface choline esterase of the axon is involved in the conduction process. Nachmansohn believes that "the release and removal of acetyl choline are intracellular processes ocurring on the neuronal surface and inseparably associated with the electrical manifestations of conduction". The significant facts supporting this concept are given in his review (page 331) (Nachmansohn and Wilson 1951). Not all of the workers in the field are in agreement with the conclusions of Nachmansohn concerning the function of the surface choline esterase of the axon (Nachmansohn 1950, 1951, and Merritt 1952). No attempt will be made here to review the pros and cons of the theory.

Surface enzymes have been implicated in acetyl choline activity in heart muscle. Welsh (1948) has reviewed a number of investigations and concludes that the effect of acetyl choline on heart muscle is due to its role as a co-factor in surface enzyme systems which are concerned with the polarity, permeability and excitability of the membrane. He quotes a study by Cook (1926) who found that methylene blue counteracts the action of acetyl choline on heart muscle. The methylene blue had to be in the medium rather than in the cells to inhibit the acetyl choline effect. For example, cells immersed in methylene blue will, after a period of time, take up considerable quantities of the dye and become deeply stained. The presence of methylene blue in the interior of the cells did not block the acetyl choline effect. Only if methylene blue was also present in the extracellular environment was this accomplished. Therefore, it was concluded that the acetyl choline effect must be at the cell surface.

Although the acetyl choline-choline esterase system is of primary importance in the functioning of the nerves and muscles, there is no concrete evidence concerning the specific mode of action. Perhaps the best clue has come from some studies of the acetyl choline-choline esterase system located in the surface of the red blood cell. Greig and co-workers, in a series of papers, have studied the effect of acetyl choline and of choline esterase inhibitors on the hemolysis of red cells suspended in buffered saline, and the parallel loss of potassium. Acetyl choline added to a red cell suspension is hydrolyzed to acetic acid plus choline. During the time that acetyl choline is being hydrolyzed, the hemolysis (Greig and Holland 1949) and the loss of potassium are both markedly decreased (Holland and Greig 1950). However, if the choline esterase is inhibited by any of a number of substances, the presence of acetyl choline does not prevent hemolysis (Fig. 5) or potassium loss (LINDVIG, GREIG and PETERSON 1951). If the cells are allowed to lose part of their potassium, the hydrolysis of acetyl choline not only prevents the further loss of this ion, but also causes a reaccumulation of the potassium previously lost (Fig. 6) (GREIG, FAULKNER, and MAYBERRY 1953). GREIG and co-workers claim that the process of hydrolysis of acetyl choline at the cell surface is intimately connected with the permeability of the membrane to ions and the ability of the cell to accumulate potassium against the concentration gradient. However, glycolysis can also activate the accumulation of potassium. GREIG suggests that the glycolysis may be necessary for production of ATP which in turn is required for the synthesis of endogenous acetyl choline, a synthesis known to occur in the red cell. In the absence of glycolysis, there is no endogenous acetyl choline, but exogenous acetyl choline can substitute.

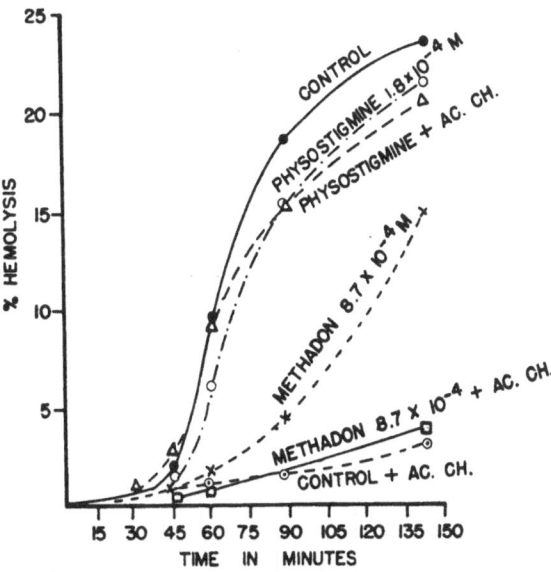

Fig. 5. The influence of choline esterase activity on hemolysis of erythrocytes.

1 cc. packed erythrocytes in saline or buffer, either alone or with acetylcholine or the cholinesterase inhibitor, or with both of these drugs, were incubated in a water bath at 37° C. Final volume 10 c. Final concentration of acetylcholine 0.01 M. Concentration of cholinesterase inhibitors is indicated on the graph. Symbols on the graph denote isotonic solutions, e. g., 90 % KCl-10 % NaCl signifies isotonic solutions of KCl and NaCl in the proportion 9:1.

(GREIG and HOLLAND 1949; courtesy Arch. Biochem.)

The studies of GREIG et al. have been criticized by PARPART and HOFFMAN (1952) who claim that the effect is not due to acetyl choline splitting per se, but to the presence of one of the products of the reaction, acetic acid. They states that the addition of acetic acid even in the presence of choline esterase inhibitors prevents the loss of potassium. At the time of this writing, PARPART's studies have only been reported in an abstract and the data are not available. GREIG, although she has never directly answered the criticism of PARPART, has claimed that the buffering was adequate in her experiments to take care of the acid production and that the pH remained constant during the course of hydrolysis. She also studied the effects of the products, acetate and choline, on the rate of hemolysis and found that they gave no protection, whereas acetyl choline did (GREIG and HOLLAND 1951). She did not report the effects of acetate or acetic acid on potassium loss. If the points of difference between PARPART and GREIG can be cleared up, and if surface choline esterase activity can indeed activate the accumulation of ions, then this might provide some of the answers to the question of the mechanism of acetyl choline action in nerve, where the establishment of a polarized state necessary for conduction is intimately associated with the ability of the nerve to transport the monovalent cations.

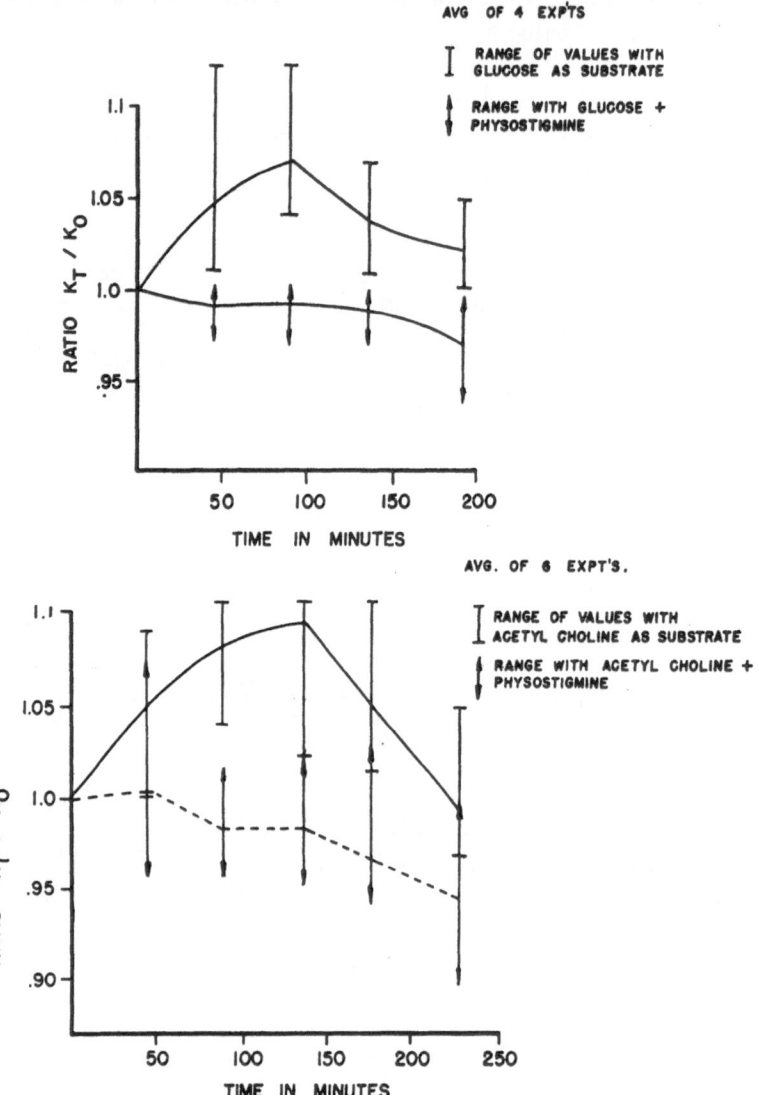

Fig. 6a—b. Effect of acetylcholine and glucose on potassium uptake by erythrocytes.
Washed human erythrocytes, from defibrinate blood which had been refrigerated for 2 days at about 5° Cl. The cells were suspended in 0.05 M sodium phosphate buffer pH 8, made isotonic with plasma by adding NaCl, and containing 35 mM KCl and either acetylcholine (0.01 M), or glucose (0.01 M) and where indicated, physostigmine ((10⁻⁴ M). Ratio of $K_T/K_0$ signifies intracellular potassium (K) at time T to that at time 0. This is plotted against time.
(Greig et al. 1953; courtesy Arch. Biochem. Biophys.)

## Proteolytic Enzymes

Recently, Morrison and Neurath (1953) have investigated the proteolytic enzymes of hemoglobin-free stroma of human red blood cells. They tested the activity using casein and hemoglobin as substrates. The stroma was found to contain considerable proteolytic activity which could not be washed away by simple extraction procedure. However, with potassium thiocyanate, an enzyme was extracted which could be activated by reducing agents as well as by zinc or ferrous ions. Two additional enzymes were obtained by a butanol extraction. One was activated by zinc but not be reducing agents. It had a pH optimum of 7.4. The second was not activated by zinc and had a pH optimum of 3.2.

## Dehydrogenases

QUASTEL (1926, 1927 a, 1927 b) was one of the first to point out that the surface of the cell might contain enzymic activity. He studied the reduction of methylene blue by resting bacterial cells in the presence of various substrates. On the basis of a comprehensive study, particularly of the kinetics of the reaction, and also of the effect of temperature and a number of other factors, QUASTEL concluded that the dehydrogenase activity responsible for methylene blue reduction must be located at the cell surface, and furthermore that these dehydrogenases were important in the metabolism of extracellular substrates.

## DPN-ase

This enzyme, which destroys the co-factor diphosphopyridine nucleotide, has been found in the red cell stroma by ALIVISATOS and DENSTEDT (1951). They observed that the lactic dehydrogenases of red cell hemolysates was more active if the stroma were removed. The inhibitory effect of stroma was shown to be due to its DPN-ase activity, which destroyed DPN necessary for the dehydrogenase activity. The function of this surface enzyme is not known.

## Ascorbic Acid Oxidase

MANDELS (1953 b) has found that spores of the fungus *Myrothecium verrucaria* actively oxidize ascorbic acid. The enzyme is readily extracted by grinding, but in the living cell it apparently is located on the surface. The evidence in favor of such a location is of two types. First, it was found that pH has the same effect on the isolated enzyme as on the enzyme activity of the intact viable spores. Since the pH of the interior of a viable cell cannot be greatly effected by the pH of the environment, and since it is unlikely that the permeability to ascorbic acid would possess the same pH curve as that for the activity of the isolated enzyme, it can be concluded that the enzyme is located in the surface, exposed to the environment. Secondly, the treatment of living spores with 0.1 N HCl for a few minutes, although it has little effect on the general cytoplasm as indicated by a normal endogenous metabolism and glucose respiration, completely inactivates the ascorbic oxidase. The enzyme must be located on the surface to be accessible to the 0.1 N HCl. If it were located inside the cell it would be protected as are the enzymes for respiration of glucose.

The function of the enzyme is unknown.

# Surface Enzymes Concerned with the Movement
# of Substances Into and Out of the Cell

The movement of substances into and out of cells has been the subject of intensive study for many years. A large variety of substances and many kinds of cells have been investigated. The concepts of permeability and active transport that have arisen from these studies are complex, and the details are beyond the scope of the present discussion. Many excellent reviews have appeared (BROOKS and BROOKS 1941; DAVSON and DANIELLI 1943; HÖBER 1946; DAVSON 1952; HEILBRUNN 1952), the most recent, in this monograph series. It is pertinent to realize that every substance passing into or out of the cell must pass through the cell surface layer by some means. There has been a growing conviction that in some cases enzymatic activity in the cell membrane may actually make possible or at least may enhance the passage of certain substances (ROSENBERG 1948; STEINBACH 1951; WILBRANDT 1954).

Two closely related questions must be considered. What is the driving force which causes substances to move into or out of the cell? By what mechanism do the substances pass through the cell membrane? In discussing these questions, it is convenient to divide the movements of substances across the cell membrane into two categories.

The first category consists of those movements of substances which are dictated by the differences in the gradient across the membrane. In its simplest terms, with non-electrolytes, a concentration difference, or more correctly, an activity gradient for a given substance is established across the cell membrane. There exists, therefore, a greater tendency for molecules of the substance to move from the side of the membrane with the higher activity to the side with the lower activity, than in the reverse direction. Consequently, if the molecules can penetrate the membrane (i. e., the membrane is permeable to that substance), there will be a net movement of the substance across the membrane which will continue until the activity gradient is abolished. The system will then be in equilibrium, with movement of the substance at equal rates in the two directions. In the case of ions, complications arise because of the electrical charge. The movement of one ion is influenced by the other ions in the system, whereas non-electrolytes behave independently of each other. With ions, both the activity gradient, and the effects of the interdependence on other ions have to be taken into account in calculating the equilibrium across the membrane. It is necessary to think in terms of the "electro-chemical gradient", rather than the "activity gradient". In either case the driving force for the net movement of the substances is *not* a function of the cell membrane, but is a function of the concentrations of substances in the cytoplasm and in the environment.

The rate at which a substance moves in response to an electrochemical gradient is determined by two factors, the magnitude of the gradient, and the resistance of the membrane to the passage of the substance through it.

This resistance is generally called permeability. The permeability of a membrane to a given substance depends on the exact mechanism by which the substance penetrates the membrane. In the earlier studies two primary factors in penetration were recognized, lipid solubility and molecular size. The membrane contains a lipid phase into which substances may dissolve and thus pass through the membrane. There may be a movement through the membrane which is restricted by pores of limited size. In addition to these factors, adsorption is thought to play a role, especially in the case of ion movements, with continual exchange of ions occurring between adjacent adsorption sites.

The mechanism of penetration listed above are relatively *non-specific*. That is, substances which are closely similar in molecular size and in physical properties, moving through the membrane by such mechanisms, should penetrate the membrane with the same ease. In recent years, it has become increasingly evident that the membrane possesses a high degree of specificity toward certain substances, a specificity incompatible with any of the classical mechanisms. This specificity has been accounted for by assuming that the substance in question can combine in a dissociable complex with a specific reactant or carrier in the membrane structure. The complex then moves across the membrane, presumably in the lipid phase and dissociates again at its inner boundary, liberating the penetrating substance. The carrier is then free to diffuse to the outer boundary to recombine with another molecule of the substance.

It is important to keep in mind that regardless of the mechanism of penetration, any net movement of a substance in response to an electrochemical gradient does not require that energy be supplied to the reaction by enzymes in the cell membrane structure. The membrane may play a completely passive, structural role. However, the cell may do work to maintain the movement of substances by maintaining the electrochemical gradient. A substance continually produced inside the cell will continue to diffuse out. A substance in the medium will continue to diffuse into the cell if it is converted to another form by reactions in the cytoplasm. In either case the enzyme reactions leading to the movements of substances occur in the cytoplasm of the cell and are not of present interest.

A second category of movements of substances through the cell membrane is that in which the electrochemical gradient is not the only direct source of energy. In such cases, the term "active transport" has generally been applied. If the substances move in a direction opposite to that dictated by the electrochemical potential, there can be no question that the cell possesses a "pumping mechanism" energized by metabolic reactions. If the movement is in the same direction as the electrochemical gradient, but proceeds at a rate which is out of all proportion to the size of the gradient and the permeability of the membrane, it is probable that a "pumping mechanism" is abetting the movement of the substance. The exact details of the "pumping mechanisms" are as yet unknown although there have been numerous theories. In many cases, it has been assumed that the pump resides in the cell membrane and that it is energized by enzymic reactions occurring in the cell surface.  It is therefore pertinent to examine some of the "active transport" systems that have been proposed, with respect to the coupled surface enzymic reactions.

## Transport of Monosaccharides

The classical studies of permeability indicated that sugars could only diffuse through cell membranes at relatively low rates as would be expected on the basis of low lipid solubility and large molecular size.  Yet in some cells certain sugars can be rapidly absorbed.  In the intestine (Barany and Sperber 1939) and in the renal tubule (Smith 1951) of vertebrates, glucose is rapidly transported against the concentration gradient.  Because the simpler mechanisms of membrane penetration cannot readily account for sugar uptake, it is entirely possible that enzymatic reactions at the cell surface may be involved.  The factors in sugar uptake have been studied rather intensively in the case of the yeast cell, the human red blood cell, the muscle cell, and also in the intestine and renal tubule.

## The Yeast Cell

Most of the studies concerned with sugar uptake in yeast have been carried out using baker's yeast, *Saccharomyces cereviseae*.  This organism can take up certain hexoses and can metabolize them rapidly under aerobic and anaerobic conditions.  In fact, the rate of metabolism is as fast as the rate of uptake, so that free sugar as such does not appear to accumulate in the cytoplasm (Berke and Rothstein 1954).  For this reason the usual methods for studying permeability and active transport can not be applied.

Baker's yeast can ferment glucose, fructose and mannose at rates amounting to 2 mM/gm. yeast (wet weight)/hour at $25^\circ$ C., with a sugar concentration as low as .05 M.  Assuming a concentration gradient of .05 M, the permeability constant for the movement of glucose through the yeast cell membrane would be $1.4 \times 10^{-9}$ cm./sec. (the average diameter of yeast cells are taken as $5 \mu$ and number of yeast cells per gram as $10^{10}$).  This figure itself is of limited usefulness because there is little data on the permeability constants of other cells to glucose.  Constants of $8 \times 10^{-12}$ have been quoted for the alga, *Chara ceratophyla*, $1 \times 10^{-11}$ for eggs of the marine annelid, *Chaeloptera*.

Despite the existence of a rapid uptake of the fermentable sugars, glucose, mannose and fructose, by yeast, other closely related sugars do not penetrate through the plasma membrane.  Conway and Downey (1950 a) found by volume of distribution techniques that galactose and arabinose pass through the cell wall, but do not distribute in the cell water to any appreciable extent in an hour (Fig. 7).  These results were confirmed (Rothstein and Meier 1954) and in addition sorbose was found to behave in the same manner.  Under the conditions of the experiment, the addition of the same amount of glucose, mannose or fructose to the yeast suspension would result in their total uptake in 3 to 5 minutes. From certain of the experiments it can be calculated that the penetration of the cell membrane by galactose, sorbose and arabinose is less than $1/_{1000}$ that by the fermentable sugars. Thus the mechanism by which sugars pass the cell membrane must be a highly specific one. The specificity does not bear any obvious relationship to the chemical and physical properties of the sugars. On the basis of molecular size, arabinose, a pentose, should penetrate faster than the hexoses. The aldoses, glucose and mannose, penetrate; the aldose, galactose, does not. The ketose, fructose, is taken up; the ketose sorbose, is not.

It seems more than a coincidence that the specificity of the cell membrane of yeast is similar to that of the yeast enzyme, hexokinase, which can phosphorylate glucose, mannose and fructose, but not galactose, sorbose or arabinose. Furthermore, if cells are given galactose in the presence of glucose, after a few hours, they develop the capacity to take up and ferment galactose, a phenomenon called enzymatic adaptation (SPIEGELMANN, REINER, and COHNBERG 1947). It is interesting that during adaptation the yeast cell membrane is altered from one that is impermeable to galactose into one through which galactose can be taken up and that at the same time, an enzyme, galactokinase, appears which is capable of phosphorylating galactose (SPIEGELMANN, REINER, and MORGAN 1947).

WERTHEIMER (1934) studied the initial events in glucose metabolism. From his data, he concluded that the cell is impermeable to glucose, but that the sugar is adsorbed on the cell surface and dehydrogenated there, to form a permeable product. CRAMER and WOODWARD (1952) also concluded that glucose must pass through the cell membrane by some enzymic reaction. They found that 2-desoxy-d-glucose competitively inhibited the fermentation of glucose by living yeast cells. However, in a cell free extract of yeast, there was almost no inhibitory effect on glucose fermentation. Thus the inhibitory effect is only apparent in the presence of cellular structure.

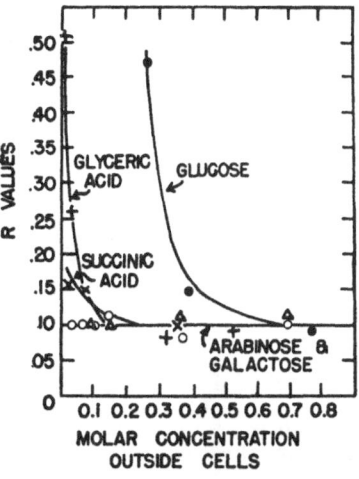

Fig. 7. The volume of distribution of various substances in the yeast cell.

The "R" value gives the amount of substance which has entered the yeast cell as a ratio of the external concentration at the time of observation (15 mins. after suspending). 1 kg. washed centrifuged yeast was suspended in 1 liter of solution. Inulin, peptone and gelatin give zero "R" value.

(CONWAY and DOWNEY 1950; courtesy Biochem. J.)

In order to study the problem further, attention was focused on the events in glucose uptake which take place on the cell surface. The studies have been reported in a series of papers from this laboratory. Because the information has been recently summarized (ROTHSTEIN 1954), only the most pertinent data will be presented here.

The first studies were carried out using uranyl ion as an inhibitor of sugar uptake.  Uranyl ion is a particularly good inhibitor because very low concentrations will block the metabolism of sugar (Booy 1940; Barron, Muntz, and Gasvoda 1948; Rothstein and Larrabee 1948), because the inhibition is specific for reactions involved in sugar uptake (Rothstein, Meier, and Hurwitz 1951) and because the uranyl ion acts on the cell surface.  Uranyl ion blocks a reaction at the cell surface concerned with the uptake of glucose and it can therefore be used as a tool to characterize the mechanism which is responsible for the passage of sugar through the surface.

The evidence for surface action of uranyl ion is of several kinds (Rothstein and Larrabee 1948).  First, it has been shown that the inhibition of sugar metabolism is associated with the binding of uranium to the cell in

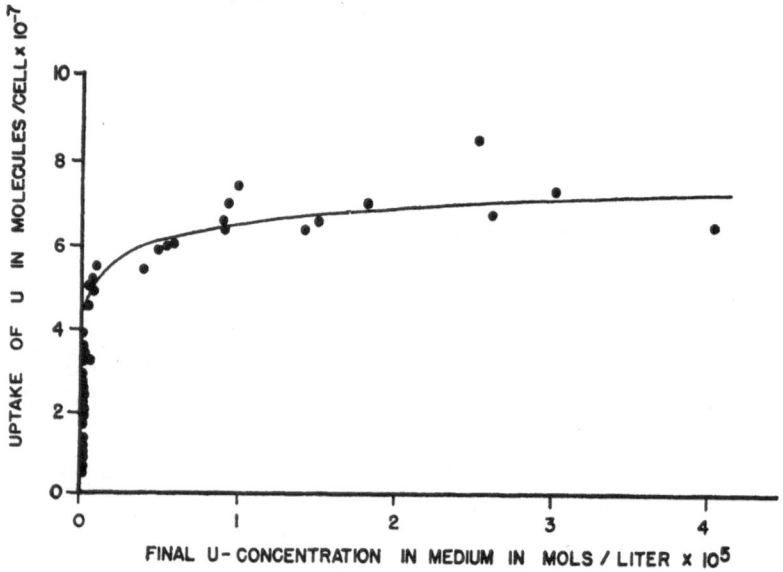

Fig. 8.  The relationship between the final concentration of $U$ in the medium and the $U$-uptake per cell after one hour. The data represent 4 experiments at yeast concentrations of 10, 20 and 40 mg./ml. (Rothstein and Larrabee 1948; courtesy J. cellul. a. comp. Physiol.)

a reversible complex.  The equilibrium between the cell and the uranium in solution is achieved at a rapid rate, in fact, so rapid that unreasonable assumptions concerning the permeability of the membrane to uranyl ion would have to be made if penetration into the interior of the cell were required for its inhibitory activity.  Second, the equilibrium between uranium and the cells can be described in terms of the mass law.  There are a specific number of binding sites which participate in the equilibrium and these can be saturated by relatively low concentrations of uranium because of the stability of the complex that is formed (Fig. 8).  Once these sites are saturated no further binding of uranium occurs.  The maximal binding of uranium amounts to a concentration in the cytoplasm of $1 \times 10^{-3}$ M per liter of cells.  Yet the general cytoplasmic contents of diffusible anions is capable of binding at least 100 times this amount of uranium. The anions of the cell are primarily bicarbonate, orthophosphate, organic phosphates, organic acids and proteins, all of which form complexes with uranium. The bicarbonate alone is almost 0.1 M and the acid-extractable orthophosphate and organic phosphate of the order of 0.02 M in each case. Thus the uranyl ion must equilibrate with sites which are not in immediate contact with the cytoplasm. Thirdly, the inhibition of fermentation is reversed by adding orthophosphate to the medium. The concentrations of orthophosphate required for marked reversal are of the order of 1 to $2 \times 10^{-4}$ M (Table 3). In view of the fact that the concen-

Table 3. *A Comparison of Required Concentration of Inorganic Phosphate Necessary to Reverse U-Inhibition with Concentration of Cellular Phosphate.*

| U Conc. | Inorganic Phosphate Added | Inhibition of Glucose Consumption | Trichloracetic Acid Extractable Cellular Phosphate | | |
|---|---|---|---|---|---|
| | | | Inorganic | Organic | Total |
| M/l. | mM/l. | % | mM/l. | mM/l. | mM/l. |
| 0 .... | 0 | 0 | 10.6 | 12.5 | 23.3 |
| $2 \times 10^{-5}$.... | 0 | 91 | 8.3 | 15.8 | 25.1 |
| $2 \times 10^{-5}$.... | 0.05 | 83 | 9.2 | 15.0 | 25.4 |
| $2 \times 10^{-5}$.... | 0.1 | 69 | 9.1 | 15.7 | 25.4 |
| $2 \times 10^{-5}$.... | 0.2 | 30 | 9.7 | 15.6 | 25.4 |

Yeast concentration—20 mg./ml.

(ROTHSTEIN and LARRABEE 1948; courtesy J. cellul. a. comp. Physiol.)

tration of orthophosphate within the cytoplasm is 1 to $2 \times 10^{-2}$ M or 100 times as high, it is evident that the uranium must be located on the outside of the cell where it is influenced by the phosphate of the medium rather than by the phosphate of the cell.

The chemical nature of the cell surface loci with which uranium combines have been investigated in some detail (ROTHSTEIN, FRENKEL, and LARRABEE 1948). It has already been pointed out that the binding of uranium is a reversible equilibrium. The simplest form of a reversible reaction can be expressed as

$$U + Y \rightleftharpoons UY$$

or, in terms of the mass law,

$$K = \frac{(U) \quad (Y)}{(UY)}$$

where $K$ is the mass law constant, $(U)$ the concentration of free uranyl ion, $(UY)$ the concentration of free yeast sites. This equation has been tested by studies of uranium-binding, and has been found to represent the situation adequately, whereas other mass law formulations with different ratios of reactants do not. The value for $K$ is in the range of 3 to $4 \times 10^{-7}$ and the number of binding sites is $1 \times 10^7$ per cell, equivalent to a concentration of $1 \times 10^{-3}$ M per liter of cells.

The chemical identity of the surface sites was established by comparing the complex formed between uranium and yeast with that formed between uranium and numerous other substances (ROTHSTEIN and MEIER 1951). Direct comparisons were made by setting up a competitive system consisting of yeast, uranium and soluble complexing agent and then measuring the distribution of uranium (Fig. 9). A wide spectrum of stabilities was found among the various complexing agents, covering a 40,000 fold range. Polyhydroxy compounds formed unstable complexes. Among the carboxyl containing compounds, monocarboxylic acids formed relatively unstable complexes. Polycarboxylic acids formed complexes of higher stability due

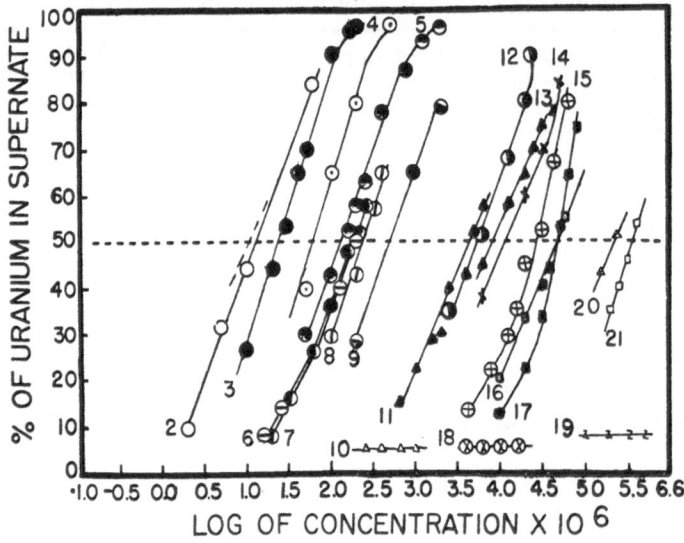

Fig. 9.  Distribution of uranium between cells and supernate in the presence of various concentrations of complexing agents.
Yeast concentration was 10 mg. wet weight per milliliter, $U$ was $5 \times 10^{-6}$ M and pH 3.5.

| | | |
|---|---|---|
| 1. Yeast | 8. Meta phosphate | 15. Orthophosphate |
| 2. Metaphosphate Polymer | 9. Nucleic acid (tech.) | 16. Maleate |
| 3. Hexameta phosphate | 10. Adenylic acid | 17. Glycerophosphate |
| 4. Desoxy ribonucleic acid | 11. Egg Albumin | 18. Glucose-1-phosphate |
| 5. Pyrophosphate | 12. Serum Albumin | 19. Glucose |
| 6. Triphosphate | 13. Citrate | 20. Acetate |
| 7. ATP | 14. HDP | 21. Fructose |

(ROTHSTEIN and MEIER 1951; courtesy J. cellul. a. comp. Physiol.)

to chelation and multiple ring closure. The proteins, which complex uranium by virtue of free carboxyl groups and associated accessory groups, formed complexes of even greater stability. However, the only compounds which formed uranium complexes with a stability as high as that formed between yeast and uranium were the multiphosphate compounds, including both the polyphosphates and nucleic acids. Within the group of multiphosphates there was a definite increase in stability of the complexes with increasing molecular weight. Highly polymerized metaphosphates most closely resembled the yeast cell in regard to binding of uranium. However, another factor must be considered. Although free ATP has only $^1/_{20}$ the affinity for uranium that the yeast cell possesses, studies of the effect of uranium on the ATP-hexokinase-Mg system (Hurwitz 1953) suggest that ATP combined with protein has a higher affinity for uranium than does free ATP. Thus it is concluded that the surface groups of yeast are either high molecular weight multiphosphate compounds or possibly ATP combined with protein.

Further evidence for the multiphosphate nature of the yeast surface groups is given by pH studies. The complexes of uranium are all markedly influenced by pH. The yeast uranium complex is influenced by pH in exactly the same manner as is the polyphosphate-uranium complex, but in an entirely different manner than is the carboxyl-uranium complex.

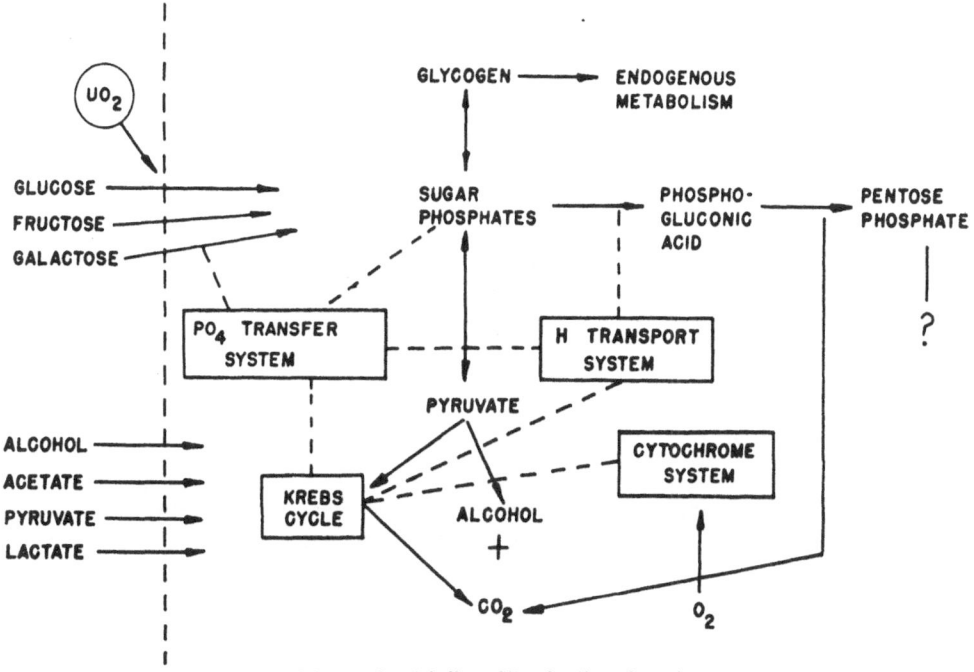

Fig. 10. Scheme of metabolism: Site of action of uranium.
(Rothstein et al. 1951; courtesy J. cellul. a. comp. Physiol.)

What is the role of the surface polyphosphate groups? The problem has been examined from several points of view. The qualitative effects of uranium are rather simple (Rothstein, Meier, and Hurwitz 1951). The uptake of all of the fermentable sugars, glucose, mannose, and fructose, is blocked under both aerobic and anaerobic conditions. If cells are adapted to galactose, then galactose uptake in such cells is also blocked. However, other forms of metabolism are not appreciably influenced by uranium concentrations sufficiently high to block sugar uptake almost completely. For example, the respiration of acetate, pyruvate, lactate and alcohol, as well as the endogenous respiration, are relatively insensitive to uranium. The synthesis of glycogen from alcohol as well as the fermentation of glycogen to alcohol (under the influence of dinitrophenol [Rothstein and Berke 1952]) are also insensitive to uranium. An examination of the accepted schemes of glycolysis and respiration leads to the conclusion that uranium must act by blocking the specific reactions which introduce the sugars into the general metabolic cycle, but that uranium has no effect on the cycle itself (Fig. 10). Two possibilities exist. First, uranium may reduce the permeability of the cell to sugar—a hypothesis suggested by Barron et al. (1948). Second, uranium may inhibit the enzymatic reaction which introduces sugars into the metabolic machinery. In this case the only known enzyme which fulfills this function is hexokinase. Implicit in the second possibility is the condition that the hexokinase is located on the surface of the cell.

Some of the properties of the cell surface reaction in sugar uptake have already been discussed, such as the high degree of specificity and the chemical nature of the uranium binding loci. Using uranium as a tool it has been possible to make other pertinent observations. For example, it has been shown that a specific number of uranium binding sites are con-

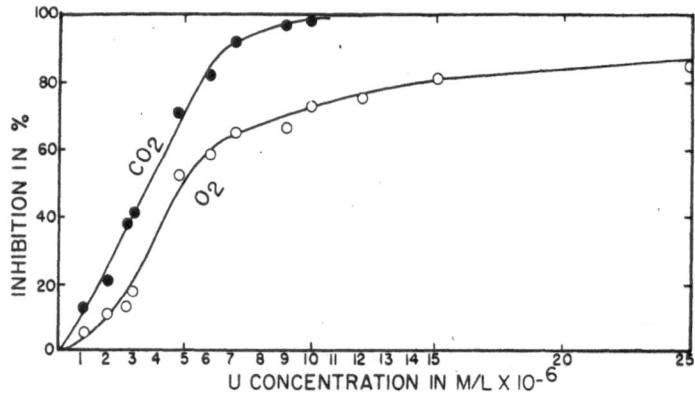

Fig. 11. The effect of uranium on the respiration and fermentation of glucose by yeast.

cerned in the fermentation of glucose and that the inhibition is directly proportional to the fraction of these sites combined with uranium (Rothstein, Frenkel, and Larrabee 1948). That is, when ¼ of the sites are combined, 25% inhibition results. When ½ are combined, 50% inhibition results. It is therefore concluded that any given site combined with uranium cannot participate in sugar uptake, but is completely blocked. In the case of respiration of glucose, the situation is more complicated (Rothstein, Meier, and Hurwitz 1951). The addition of sufficient uranium to block all of the sites involved in fermentation leads to only a 60% inhibition of respiration. In order to block the last 40% of the respiration, the uranium concentration must be increased some 15 fold (Fig. 11). Thus the inhibition of respiration involves two chemically distinct kinds of surface loci. The first of these is saturated by low uranium concentrations. It forms a very stable complex with uranium and has been identified as polyphosphate. The second is saturated only in the presence of relatively high uranium concentrations and the complex formed is therefore considerably less stable than that with polyphosphates. Although the chemical nature of the second type of surface site has not been characterized in detail, preliminary studies suggest that protein carboxyl groups are involved.

It has been pointed out previously that the yeast cell membrane can differentiate between glucose, fructose and mannose as compared with galactose, sorbose and arabinose, the latter sugars being able to penetrate only slowly if at all. Studies of uranium inhibition indicate that other kinds of differentiation are also built into the membrane. Glucose uptake under anaerobic conditions involves primarily surface sites identified as polyphosphates whereas glucose uptake under aerobic conditions involves apparently independent kinds of surface sites, one identified as polyphosphate and a second as protein carboxyl. Another surface differentiation has been found in the case of galactose-adapted cells. The uptake of galactose by such cells is more sensitive to uranium than is the uptake of the other sugars, indicating that fewer surface sites are involved in the case of galactose. Hence the organization of that part of the membrane responsible for sugar uptake is complex. The membrane possesses inherent specificity for certain sugars. It also possesses at least two chemically distinct kinds of anionic groups capable of binding uranium, one a polyphosphate and one perhaps a protein carboxyl. Both kinds of sites are independently functional in aerobic uptake of glucose, but fermentation involves primarily polyphosphate sites. Different numbers of fermentation sites are functional in galactose uptake in adapted cells than in uptake of the other sugars.

How do the uranium binding sites function in sugar uptake? It has already been shown that the rate of sugar uptake is dependent on the number of functional surface sites (those which are not complexed with uranium). Further light is thrown on this question by kinetic studies. HOPKINS and ROBERTS (1935) have shown that fermentation of glucose follows the Michaelis-Menten equation:

$$\frac{1}{V} = \frac{Km}{Vm} \cdot \frac{1}{S} + \frac{1}{Vm}$$

where $V$ is the rate of metabolism, $S$ the substrate concentration, $Vm$ the maximal rate, and $Km$ the Michaelis constant. The equation is predicated on the existence of a substrate enzyme complex, the concentration of which determines the rate of the reaction. Thus,

$$E + S \rightleftharpoons ES \rightarrow \text{product.}$$

Inherent in the equation is the concept of a limited number of enzyme sites, so that with increasing substrate concentration, the rate reaches a maximal value associated with saturation of the enzyme sites with substrate.

The fact that normal fermentation is fitted by the Michealis-Menten equation may only be a reflection of the rapid penetration of glucose into the cell followed by a slower process, its metabolism. Thus the limiting factor would be an internal enzyme which would give rise to the Michaelis kinetics. However, if penetration of glucose were more rapid than its metabolism, then free glucose should accumulate in the cell and this is not the case (Berke and Rothstein 1954). Furthermore, if a cell surface reaction is made the rate limiting step by inhibiting with uranium, the anaerobic uptake of glucose still can be characterized by the Michaelis-Menten equation (Fig. 12) (Hurwitz and Rothstein 1951). Therefore, it must be concluded that an interaction between glucose and the cell occurs at the surface which involves a saturation phenomenon. Glucose combines with some constituent of the surface which is present in limited concentrations.

The type of inhibition is non-competitive as indicated by a constant value for the $Km$ at different uranium concentrations, but altered value for

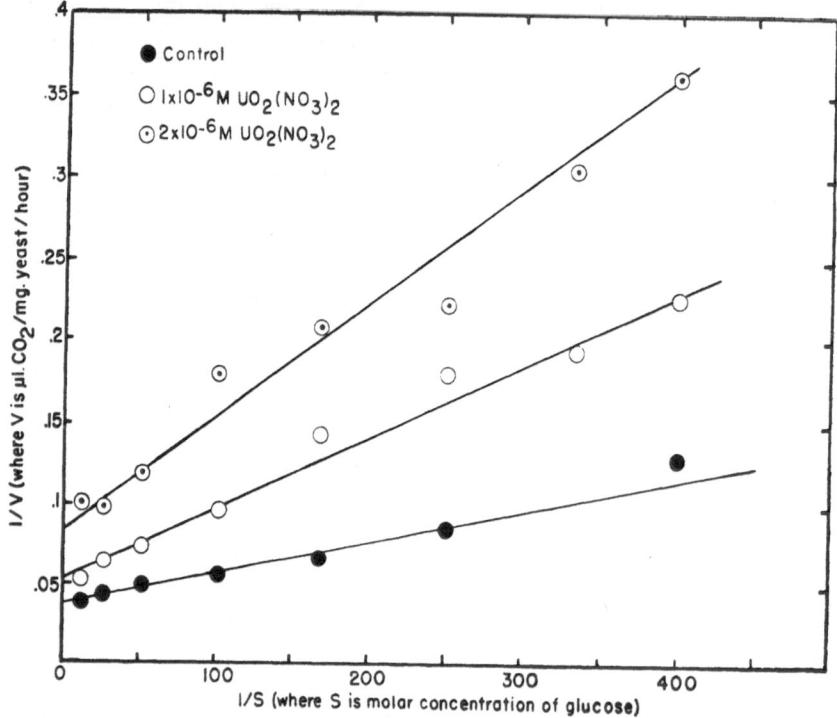

Fig. 12.  Kinetics of inhibition of fermentation of glucose.
Substrate concentrations ($S$) are expressed in M/l.  Rates ($V$) are in µl. of $CO_2$ per mg. yeast (wet weight) per hour. The yeast concentration was 5.15 mg./ml., the temperature 25° C., and the pH 3.5.
(Hurwitz and Rothstein 1951; courtesy J. cellul. a. comp. Physiol.)

the $Vm$. This suggests that uranium and glucose combine with different loci on the surface sites (otherwise competition would result), but that any site complexed with uranium, although it may still be able to combine with glucose, can no longer carry out the necessary reaction.

The kinetics of inhibition of respiration are more complicated. At certain uranium concentrations the data do not fit the Michaelis-Menten equation, at others they do. However, the values of the constants, $Km$ and $Vm$, do not fit any simple relationship. The complexity of the kinetics under aerobic conditions is consistent with the previously indicated finding that two independent mechanisms of uptake are operative.

The effect of temperature on the rate of uptake of sugars has been measured in the presence and absence of inhibiting concentrations of glucose (HURWITZ and ROTHSTEIN 1951). In each case the energy of activation is high, of the order of 22,000 calories per mol, a value which indicates a considerable energy barrier which must be overcome in the penetration of glucose through the membrane.

The nature of the surface reactions has been studied not only in terms of their inhibition by uranium but also in terms of their stimulation by extracellular ions. The connection of uranium inhibition with polyphosphate groups and the possibility that hexokinase activity may be involved suggests that uranium may act by displacing the bivalent cation co-factors in phosphorylation reactions. The first studies were concerned with the competition of various cations with uranium in regard to binding by the yeast cell. It was found that appropriate concentrations of Mg, Ca, Ba and Zn added to a suspension of yeast plus uranium displaced uranium from the yeast cell (ROTHSTEIN and MEIER 1951). However, relatively high concentrations of the competing ions were required. It could be calculated that the uranyl ion possessed a greater affinity for yeast sites by a factor of about 5000. Tests of competition between uranium and K or Na were negative at the concentrations used.

The interaction between physiologically active bivalent cations Mg, Ca and Mn and the yeast cell was investigated in more detail, using isotopic techniques with $Mn^{54}$ and $Ca^{35}$ (ROTHSTEIN and HAYES 1954). Mg was studied by its competition with $Mn^{54}$. The yeast cell behaves in many ways like a particle of exchange resin insofar as the bivalent cations are concerned. There are a discrete number of binding sites for Mn, Ca and Mg, amounting to $1 \times 10^{-3}$ M/liter of cells, the same as for uranium. The dissociation constants for the complexes are of the order of $1 \times 10^{-3}$, a figure which is only approximate because of the long extrapolation involved in its computation. Thus the complexes of Mn, Mg and Ca are less stable than that of uranium by a factor of about 5000, a value which agrees with that obtained by direct competition experiments with uranium. It is evident, therefore, that there is an identity between the binding sites for uranium and for the other ions. Thus the binding of Mn, Ca and Mg, as is the binding of uranium, must be on the surface of the cell. Any metabolic effects of these ions must be associated with reactions at the cell surface.

Additional evidence for the surface binding of the bivalent cations is available. The exchange equilibrium between the cell and the ions is attained as rapidly as measurements can be made (less than two minutes). Thereafter no further uptake of the ions is measureable in the next hour. The total bivalent ion content of the cell is of the order of .03 M/liter of cells (Mg .027 M, Ca .0025, and Mn .0003). At equilibrium, the *maximum* exchange amounts to .001 M/liter of cells. Thus, at most only 3% of the bivalent ions of the cells can exchange, the rest being inaccessible. If the extracellular cations were able to penetrate into the cytoplasm, it would be expected that they should be able to exchange with a far larger proportion than 3% of the bivalent ion content of the cell.

The exchange equilibrium between the cell and bivalent ions is not influenced by the metabolism of the cell.  It is the same in resting cells as in cells fermenting or respiring glucose.  However, if phosphate is present during the metabolism of glucose, phosphate is transported into the cell

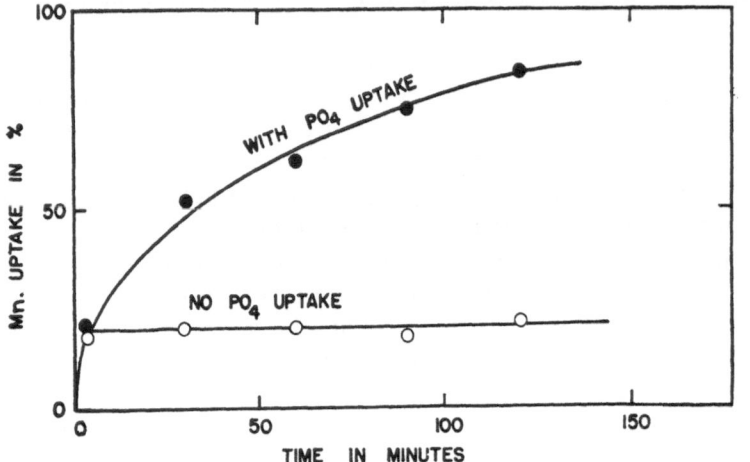

Fig. 13.  $Mn^{54}$ uptake by yeast cells in the presence and absence of a phosphate uptake.
The yeast concentration was 100 mg. (wet weight) per ml. of suspension, buffered at pH 4.5 with triethyl-amine (TEA) tartrate, succinate buffer.  $MnCl_2$ concentration was $7 \times 10^{-4}$; glucose 0.1 M and KCl 0.02 M.
Phosphate uptake was obtained by addition of phosphate in a concentration of 0.02 M.

and bivalent ions if present in the medium are carried into the cells as a phosphate complex (Schmidt, Hecht, and Thanhauser 1949).  The uptake of the bivalent ions in association with phosphate uptake is entirely

Table 4.  *The Uptake and Back Exchange of $Mn^{54}$ by Yeast.*
Yeast was suspended in distilled water at pH 3.5 in a concentration of 200 mg. (wet weight) per ml. of suspension.  The same amount of $Mn^{54}$ was added to each suspension.  The total Mn concentration was adjusted with $MnCl_2$.

| Total $MnCl_2$ Concentration M/l. | $Mn^{54}$ Uptake in % | |
|---|---|---|
| | 2 min. | 45 min. |
| $0.4 \times 10^{-4}$ .................................... | 53 | 52 |
| $7.2 \times 10^{-4}$ ................................. | 22 | 24 |
| $0.4 \times 10^{-4}$ ................................. | 51 | 52 |
| As above, then switched to | | |
| $7.2 \times 10^{-4}$ ................................. | 27 | 26 |
| $7.2 \times 10^{-4}$ plus K, $PO_4$ and glucose ....... | 28 | 100 |
| As above for 45 min., then resuspended in | | |
| $7.2 \times 10^{-4}$ ................................. | 97 | 96 |

different from the exchange equilibrium previously described.  The uptake in association with phosphate is not an equilibrium (Rothstein and Hayes 1954).  It continues as long as phosphate uptake continues (Fig. 13).  In the presence of an excess of phosphate all of the bivalent ions in the medium are carried into the cell. The total uptake has amounted to more than $2 \times 10^{-2}$ M/liter of cells in some experiments, thus increasing the cellular content of these ions by 80%. In contrast, the exchange equilibrium amounts to only $1 \times 10^{-3}$ M/liter of cells. Bivalent cations taken up in association with phosphate are no longer exchangeable with the bivalent ions of the medium, whereas the bivalent ions taken up by exchange equilibrium are completely exchangeable (Table 4). It seems apparent then, that the bivalent ions of the cytoplasm are not accessible for exchange with those in the medium, and they can be differentiated from those bound by cell surface groups on this basis.

What is the metabolic function of the surface bound bivalent ions? In the first place, calcium and magnesium can reverse the inhibition of uranium by displacing the latter ion from the surface sites. Glucose uptake cannot proceed when the surface sites are complexed with uranium, but can proceed when the same sites are complexed with calcium and magnesium. An important question still remains, however. Can glucose be taken up if the surface sites are not complexed with any bivalent ion? This question has been difficult to answer unequivocally. It is difficult to wash the bivalent ions away from the surface sites without drastic treatment, just as it is difficult to wash them away from the surface of an exchange resin. In the case of yeast the situation is complicated by the presence in the cytoplasm of a large reservoir of cations which equilibrate slowly with the surface sites.

If bivalent ions are added to well washed cells, the effect is dependent on the extracellular pH. At pH 5.0 and below, there is a definite stimulation of the uptake of glucose with Mg, Ca or Mn, of the order of 20%. At pH 6.0, where the normal rate of fermentation is considerably reduced (see following section on the effects of pH on the rate of fermen-

Table 5. *The Effect of Bivalent Ions on the Rate of Fermentation of Glucose at Different pH's.*

Rates are expressed as $\mu$ l. of $CO_2$ per mg. of yeast (wet weight) per hour. Ion concentrations were .01 M. Buffer triethylamine (TEA), succinate tartrate at pH 3.5 and 6.0, and trishydroxyaminomethane (THAM) at pH 8.5. Glucose concentration was 0.1 M.

|  | pH 3.5 | pH 6.0 | pH 8.5 |
|---|---|---|---|
| Control | 26 | 14 | 39 |
| TEA Cl. | — | 14 | — |
| Ca | 33 | 31 | 33 |
| Mg | 30 | 27 | 38 |
| Mn | 30 | 28 | — |

tation), the stimulation by the bivalent ions is considerable, of the order of 80 to 100%. At pH 8.0 to 9.0, Mg and Mn have no effect on the rate of sugar uptake, but higher concentrations of Ca have definite inhibitory effects, of the order of 30 to 40% (Table 5).

Attempts have been made to deionize the surface sites of the cell of bivalent cations by treatment with cation exchange resins. Resin treatment for a few hours reduces the ability of the cells to take up glucose, by 30 to 40%. With overnight treatment, the cells can no longer ferment. However, the interpretation is complicated. A part of the inhibition can be reversed by adding bivalent cations to the medium of such resin treated cells. Another part of the inhibition is associated with the loss of potassium, and still another part is not reversed by replacing ions and may be

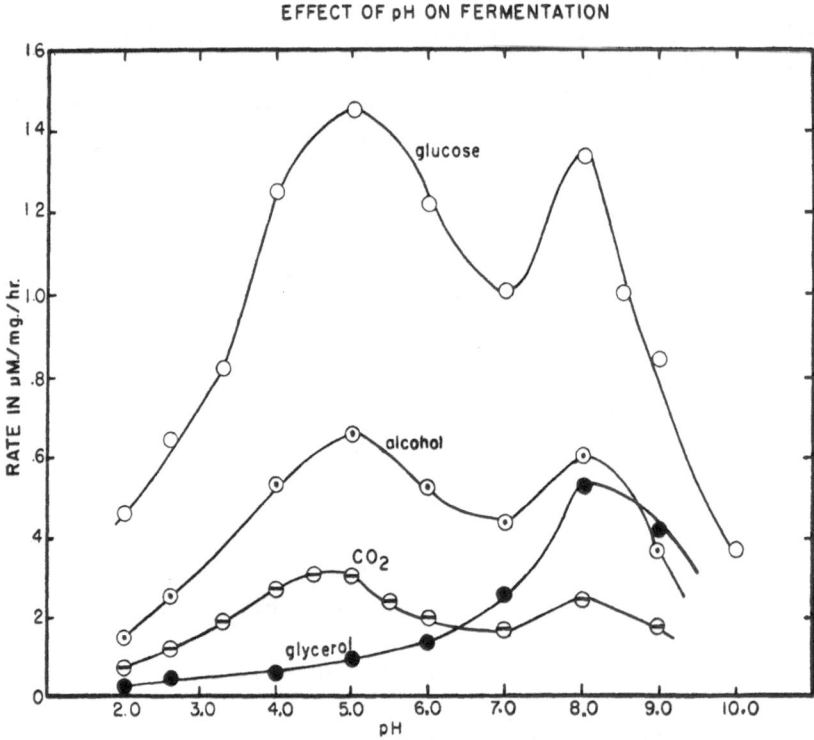

Fig. 14. The influence of pH on the uptake of glucose and on the production of alcohol, glycerol, and $CO_2$. Washed yeast was suspended in a TEA-tartrate-succinate buffer (pH 2.0 to 6.0), TRIS buffer (pH 8.0 to 10.0) and no buffer at pH 7.0 (pH was maintained by constant addition of TEA).

associated with the loss of intracellular ions. It can be concluded that the uptake of glucose is definitely influenced by the presence of bivalent cations on the surface sites, but the exact qualitative and quantitative relationship have not been determined because of technical difficulties. It is not known whether there is an absolute dependence of glucose uptake on the presence of surface bound bivalent ions.

Extracellular monovalent ions such as H+, K+ and $NH_4^+$ also can markedly influence cell surface reactions in glucose uptake. It has been accepted in the past that fermentation is relatively independent of extracellular pH over a very wide range. However, if monovalent cations. particularly K and $NH_4$, are excluded from the buffer system, fermentation is found to be remarkably dependent on extracellular pH (ROTHSTEIN and DEMIS, 1954). In order to exclude alkaline earth cations, buffer systems containing biologically inert organic cations such as triethylamine (TEA) and trishydroxyaminomethane (THAM) were used. The pH dependence of glucose uptake was then found to follow a biphasic curve with optima at 4.5 and 8.0, with a sharp dip at pH 7.0 (Fig. 14). The sharpness of the curves could be accentuated by prolonged starvation and washing, and also by treatment with a TEA-cation exchange resin (DOWEX 50). In each case the effect was found to be associated with the removal of extracellular and surface potassium (see following discussion on potassium).

The pH dependence of glucose uptake is associated with the effects of $H^+$ on cell surface reactions. This reasoning is supported by evidence that the intracellular pH during fermentation is constant and independent of the extracellular pH. CONWAY and DOWNEY (1950 b) found that during active fermentation the extracellular pH drops rapidly, but the intracellular pH rises from 5.8 to 6.2 and remains at this value. ROTHSTEIN and DEMIS (1954) found that during fermentation with the extracellular pH fixed at values ranging from 2.0 to 10.0, the intracellular pH remained constant at pH 6.2. Thus the dramatic effects of extracellular pH on the rate of fermentation must be associated with reactions on the surface exposed to the varying extracellular pH rather than the constant intracellular pH. The constancy of the internal pH is also indicated by the fact that the endogenous respiration (see also BARRON, ARDAO, and HEARON 1950) and respiration of alcohol are not influenced by extracellular pH over a wide range. The enzyme system involved are pretected by the constant cytoplasmic pH.

In addition to the effects on the rates of metabolism, the extracellular pH can influence the end products of fermentation (ROTHSTEIN and DEMIS 1954). During fermentation in an acid environment the end products are primarily $CO_2$, alcohol and polysaccharide reserves. During fermentation in an alkaline environment, less polysaccharide is formed, but large amounts of glycerol are produced. The pH curve for glycerol production is parallel to the alkaline phase of sugar uptake (Fig. 14). Thus the surface reactions are not simply a means of delivering glucose into the cell for disposition there, but they actually determine to some extent the nature of the end products.

It has been indicated previously that the effects of $H^+$ on sugar uptake are only seen in the absence of $K^+$. This is due to the fact that there is a competition between $K^+$ and $H^+$ (ROTHSTEIN and DEMIS 1953). The latter ion represses the rate of sugar uptake and the former can counteract this effect. For example, at pH 4.5, the addition of potassium stimulates the rate of glucose uptake only slightly. However, as the pH is reduced, the rate decreases markedly. The addition of an appropriate concentration of $K^+$ at any pH returns the rate to the maximal level. The lower pH, the greater the stimulation, but the higher the K concentration necessary to achieve the effect. Optimal rates can be maintained at any pH tested below pH 4.5, if the ration of K to H is maintained at about 10 to 1 (Fig. 15).

It has already been indicated that the H⁺-effect is on cell surface reactions. Thus, K⁺, which counteracts the H⁺-effect, is also acting on the cell surface. In addition, independent evidence is available that the effect of K⁺ is on the surface. The stimulating action of K⁺ can be invoked under experimental conditions such that no net uptake of K⁺ by the cells occurs. Furthermore, it has been found that the stimulating effect is independent of the intracellular K⁺ content, but is completely dependent on the extracellular K⁺ concentration.

Comparisons have been made of the effects of the alkaline earth metals. K invokes the greatest stimulation. Rb somewhat less, and Na, Li and Cs, a smaller but definite effect. None of the other ions inhibit the K effect.

Potassium stimulates not only below pH 5.0, but also in the pH range 5.5 to 7.5, where a marked dip in the pH activity curve of glucose uptake occurs. In the presence of appropriate K⁺ concentrations the biphasic pH curve is obliterated and there is instead a wide plateau from pH 2.0 to 9.0. The effect of K⁺ in the pH range, 5.5 to 7.5, differs from that in the range, 2.0 to 4.0, in two respects (Rothstein and Demis 1954). In the first place, NH₄⁺ will duplicate the K effect in the pH range 5.5 to 7.5, but has almost no effect in the acid range (Table 6). In the second place, the K⁺ effect in the neutral range is associated with a marked increase in glycerol production, whereas that in the acid range is not.

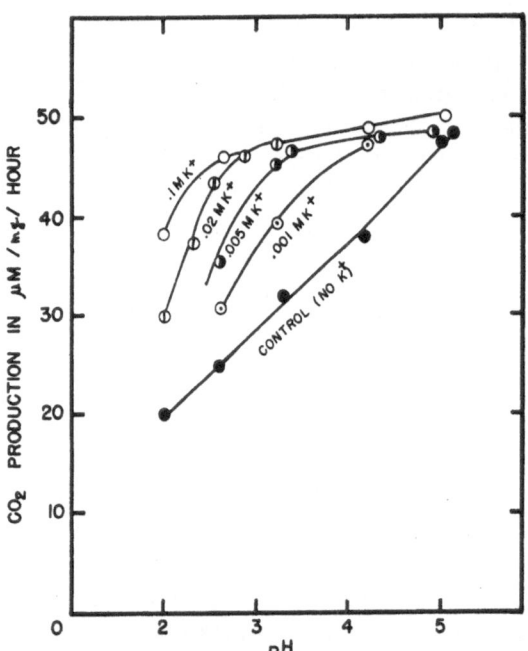

Fig. 15. The effect of the potassium concentration on the rate of fermentation at different pH values. (Rothstein and Demis 1953; courtesy Arch. Biochem. Biophys.)

In considering the nature of the cell surface reactions involved in glucose uptake, all of their properties must be taken into account. These can be listed as follows:

(1) Specificity for glucose, mannose and fructose with no uptake of galactose, sorbose, or arabinose.

(2) Uptake of galatose only in adapted cells associated with the appearance of the specific enzyme galactokinase.

(3) Surface reactions inhibited by uranium are specifically associated only with the reactions introducing sugars into the metabolic scheme.

(4) Association of the surface reaction in fermentation with a specific number of polyphosphate sites in a first order relationship as shown by uranium binding and inhibition studies.

(5) Association of 40% of the aerobic uptake of glucose with additional surface sites chemically distinct from polyphosphates, probably protein carboxyl groups.

(6) The kinetics of the surface reactions in fermentation obey a typical non-competitive inhibition.

(7) Temperature characteristic of the surface reaction is high—of the order of 20,000 cal/mol.

Table 6. *The Effect of $NH_4^+$ and $K^+$ on the Rate of Fermentation of Glucose at Different pH's.*
Rates are expressed as $\mu$ l. of $CO_2$ per mg. of yeast (wet weight) per hour. Ion concentrations were .03 M. Buffer, triethylamine (TEA) succinate, tartrate. Glucose, 0.1 M.

|  | pH 2.6 | pH 5.0 | pH 6.0 |
|---|---|---|---|
| Control ......................... | 14.3 | 43.9 | 22.7 |
| $NH_4^+$ .......................... | 15.3 | 55.4 | 50.2 |
| $K^+$ ............................. | 54.1 | 55.9 | 47.6 |

(8) Surface reactions are markedly influenced by pH, giving optima at 4.5 and 8.0.

(9) Surface reactions determine the nature of some of the end products of fermentation.

(10) Bivalent ions such as $Ca^{++}$, $Mg^{++}$ and $Mn^{++}$ combine with surface sites resulting in a stimulation of glucose uptake in the acid pH range and in inhibition in the case of Ca at alkaline pH.

(11) $K^+$ can markedly stimulate surface reactions on the acid side of the pH optima by a competitive reaction with $H^+$.

In reviewing the problem of sugar uptake in yeast, ROTHSTEIN (1954) has suggested that the various properties of the initial interaction at the cell surface exclude the possibility of any non-specific mechanism of penetration. Furthermore, the data are not consistent with the hypothesis of a specific non-enzymic carrier system. For example, how could a carrier system which simply delivers glucose to the interior of the cell, as unaltered glucose, in any way alter the nature of the end products of the reaction in response to an extracellular factor such as pH? Other evidence which does not favor the carrier system can be cited. Yeast cells are impermeable to galactose, yet they can be "adapted" to utilize galactose by proper exposure to this sugar. On the basis of a carrier mechanism it would have to be postulated that a specific carrier for galactose is produced during adaption *in addition* to the new enzyme, galactokinase, which is known to appear. Furthermore, two kinds of glucose carrier would have to be available, one of which is operative only under aerobic conditions.

The hypothesis that surface enzyme reactions are necessary for sugar uptake is much more attractive. It is compatible with all of the data, the sugar specificity, the requirement for polyphosphate, the effects of $K^+$, $H^+$, $Ca^{++}$, $Mg^{++}$ and $UO_2^{++}$, the aerobic versus anaerobic effects, kinetics, etc. What specific enzymes might be involved? The only known phosphorylating enzyme which uses glucose as a substrate is the hexokinase crystalized from yeast by Kunitz and McDonald (1946) and by Berger, Slein, Colowick, and Cori (1946). The properties of the surface reactions of yeast in the alkaline range are very similar to those of the hexokinase: (a) the pH optimum of 8.0 to 8.5 is the same in both cases, (b) both are inhibited by $Ca^{++}$, (c) the sugar specificity is the same, (d) the ratio of activities toward glucose and mannose is the same (2/1), (e) both are stimulated by $K^+$ on the acid side of the pH optimum (unpublished observations in the case of crystalline hexokinase), and (f) both are inhibited by tripolyphosphate (Vishniac 1950), although this substance cannot penetrate the intact cell (Rothstein and Meier (1948).

The properties of glucose uptake by yeast in the acid range cannot be accounted for by the properties of the crystalline hexokinase. The enzyme has little or no activity in the acid range. It was therefore postulated (Rothstein 1954) that yeast possesses a second hexokinase with an acid pH optimum (4.5). Preliminary attempts to isolate such an enzyme have met with some success (Rothstein and Bruce 1954). A soluble enzyme has been obtained from yeast, which will phosphorylate glucose in the presence of ATP and Mg and which has an acid pH optimum of about 4.0. In the presence of $K^+$ it has marked activity at a pH as low as 3,0. Quantitative determinations of activities under different conditions have been exceedingly difficult because of the instability of the enzyme. Unlike the crystalline enzyme of Kunitz and McDonald (1946) and Berger, Slein, Colowick, and Cori (1946), the new enzyme is inactivated rather than protected by cysteine. All attempts at stabilization have thus far been unsuccessful. About 90% of the activity is lost in one hour in the cold. Thus each fresh preparation will suffice for only one or at the most two determinations. Nevertheless, the preliminary data indicate that the new enzyme is a second hexokinase with an acid pH optimum, and properties similar to those of the surface reactions of yeast in an acid environment.

The properties of the surface reactions in sugar uptake by yeast are compatible with the concept that two hexokinases are located on the cell surface. Depending on the extracellular pH and $K^+$ content, one or both are active in phosphorylating glucose. The surface systems must, however, be more complicated than this. The hexokinase reaction uses up ATP which must be replenished. Furthermore, the product, glucose-6-phosphate, does not escape into the medium in measureable quantity, nor can glucose-6-phosphate added to the medium be metabolized by the cell. Thus a barrier, presumably anionic in character must be located between the hexokinase and the medium, which prevents the inflow and outflow of phosphorylated compounds, but does not prevent the inflow of glucose or of the various cations which influence the reaction, such as $H^+$, $K^+$ and $UO_2^{++}$.

In regard to the resynthesis of ATP for use by the surface hexokinase, experiments with a cell free preparation throw some light on the problem (ROTHSTEIN, DEMIS, and BRUCE 1954). BAKER's yeast is air dried at room temperature, then lyophilized and pulverized. Such cells are dead. They will no longer grow or divide. The cell membrane is destroyed as an effective permeability barrier. All of the potassium and soluble phosphates leak out, as well as considerable protein, including enzymes such as carboxylase. In fact, over 40% of the dry weight of the yeast can be washed away. Nevertheless, the residual insoluble material of the yeast cell, if supplied with adequate K+, can ferment glucose at a rate as high as 80% of that of the original living yeast cell. Furthermore, it absorbs and esterifies inorganic phosphate. It will not respire glucose or other substrates and possesses almost no Pasteur effect. The fermentation is complete, the end products being alcohol and glycogen. Thus all of the enzymes of fermentation are present in this insoluble residue. Yet the enzymes themselves are all *soluble*. Apparently the fermentative enzymes in the intact cell are associated with an insoluble element of the cell. They are rendered soluble only after autolysis and proteolysis, or after drastic treatment such as freezing and thaving, grinding or homogenizing, treatments which break up the structural properties of the cell.

Other evidence suggests not only that the fermentation enzymes are associated with an insoluble structure, but also that this constitutes a definite organelle of the cell, with its own permeability properties. If glucose plus inorganic phosphate is added to the preparation, the glucose is fermented and some of the inorganic phosphate is esterified, forming the various phosphorylated intermediates of sugar fermentation. However, these intermediates do not appear in the medium. They are formed in the insoluble fraction and do not leak out. They can only be removed by drastic treatment such as trichloracetic acid extraction. Furthermore, these same phosphorylated compounds, when added as substrates, cannot be utilized. The structure is impermeable to their movement in either direction, but nevertheless will ferment glucose, exactly as is the case in the intact yeast cell. On the basis of these studies, it is concluded that the fermentative enzymes of the cell are associated with a sub-cellular element, which has been called, for convenience, the "glycosome."

The glycosome occupies a maximum of some 7 to 8% of the original cell volume as determined by volume of distribution techniques using glucose phosphates which cannot penetrate the structure. It has already been pointed out that the "glycosome" possesses properties attributed to the surface of the cell in respect to its impermeability to phosphorylated compounds. In preliminary studies, it has also been found to behave in a similar manner with respect to the action of H+ and K+. Furthermore, it has already been pointed out that if hexokinase is present at the cell surface, some means of replenishing the ATP used up in phosphorylating glucose must be available. The only reactions in fermentation capable of producing ATP are those associated with diphospho-glyceraldehyde and with phosphopyruvate. Thus, from a biochemical point of view, if hexokinase on the surface of the cell, much of the rest of the fermentative scheme must also be at the surface. It is therefore suggested that the "glycosome", containing the fermentative machinery of the cell, lies at the cell surface. Studies of phosphate uptake support this view (see later discussion). The "glycosome" itself may not necessarily constitute the permeability barrier of the cell, for it remains intact after the permeability barrier is destroyed. The thickness of the glycosome, based on its maximum volume of 7 to 8% of the cell space, would be about 0.5 $\mu$. Thus it probably constitutes a "crust" or cortical layer around the outside of the cell. It may constitute the gelatinous cytological zone around the outside of the cell known as the cortex.

It is of some interest that the hexokinase of brain and other tissues, as in yeast, is associated with a particulate fraction (Crane and Sols 1953).

### The Human Red Blood Cell

The red blood cell of man (and other primates) possesses a unique property, a high degree of permeability toward certain sugars, whereas the red cells of other animals can be characterized as being relatively impermeable to sugars. Furthermore, the movement of glucose into and out of the human red blood cells shows certain peculiarities which suggest that a special mechanism exists in the cell surface which actively transports or at least assists the sugar across the membrane. Whether or not the mechanism is itself enzymic in nature, or is associated with cell surface enzymes has not been established with complete certainty but the evidence certainly suggests the possibility that surface enzymes are involved in some manner. For a discussion of the earlier work on uptake of sugars by red cells the reader is referred to the reviews of LeFevre (1954) and of Wilbrandt (1954) who have independently elucidated the mechanisms involved in some detail. The evidence for a surface mechanism involves studies of the kinetics of the sugar movements, the specificity and mutual interference in mixtures of sugars, and the action of inhibitory substances.

Sugar added to a red blood cell suspension in low concentrations equilibrates between the cells and medium. Although the rate of equilibrium is unexpectedly rapid, the equilibrium itself is the predicted one on the basis of a diffusion phenomenon. However, the rate of equilibration, as a function of glucose concentration, deviates from diffusion laws. At higher glucose concentrations, the rate reaches a maximum, indicating the saturation of a component (Fig. 16). At relatively high glucose concentrations, not only is the approach to equilibrium delayed, but the movement of glucose may be almost completely blocked.

When a number of different sugars are studied, it is found that the ketoses penetrate at a considerably lower rate than the aldoses and, furthermore, the kinetics of the movement of ketoses fit reasonably well the predictions of Fick's law of diffusion. All of the aldoses, on the other hand, show the saturation phenomenon already described for glucose. When pairs of sugars are used, keto or aldo, they interfere with each other as though a common reaction were involved for all. Further evidence of a common carrier is found in the common sensitivity to inhibitory agents.

A number of substances inhibit the movements of glucose into and out of the red cell. Uptake is inhibited by Hg++, p-chloromercuribenzoate, chloropicrin, bromacetophenone, allyl mustard oil, and gold, but not by Cu++, alloxan, mapharsan, iodoacetate, or arsenite. The glucoside, phlorizin, and especially its aglucone, phloretin, also inhibit sugar transport but apparently only in the outward direction. Recently WILBRANDT has shown

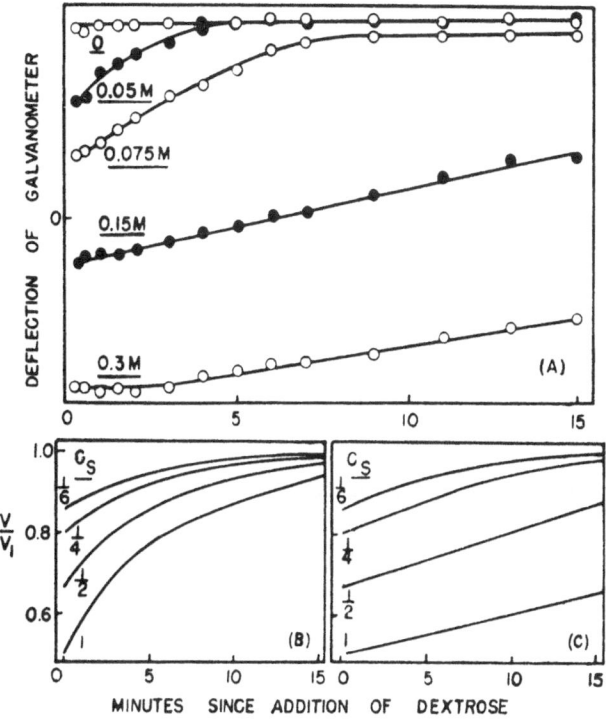

Fig. 16.   Kinetics of swelling in glucose-saline solutions.

(A) At zero time, 1 ml. saline medium with dextrose at 6 times final concentration indicated added to 5 ml. cell suspension in medium. Deflections plotted as movement from position prior to addition of glycerol; upward deflection indicates swelling of cells. Difference between initial and terminal levels results also from dilution of suspension.

(B) Predicted relations in similar experiment, assuming passive diffusion, with $dS/dt = (C_S - S/V)$, with $k = 0.2$ iso-volumes per minute.

(C) Predicted relations in similar experiment, assuming diffusion rate is limited to maximal value, $m$, ($=0.02$ iso-content per minute) by process involving cellular component.

(LE FEVRE 1948; courtesy J. gen. Physiol.)

that phloretin phosphate also inhibits the outward movement, but in this case it was possible to show that the inhibitor was not penetrating into the cell, but was acting on some surface mechanism.

Both WILBRANDT and LEFEVRE are in essential agreement concerning the properties of the system involved in glucose transport in the human red cell, but they have made individual interpretations of the data. Both agree that a transport system must exist in the membrane of the cell which involves as an essential step the combination of the sugar with a carrier

substance.  LeFevre has developed the thesis that a single carrier substance combines with glucose and the other sugars at one side of the cell
membrane in a reversible equilibrium.  The complex diffuses across the
membrane and dissociates by a reversible equilibrium at the other side.
He has developed equations, based on the mass law constants for the
formation and dissociation of the glucose carrier complex, that will fit the
kinetic data on sugar movements singly or in mixtures except at glucose
concentrations approaching isosmotic.  LeFevre suggests that high concentrations of glucose somehow block the carrier reactions.  He believes
that "there is no evidence that the red cell is equipped with a hexose
'pump' that can provide the energy for transporting sugar against a concentration gradient." He suggests, however, that the carrier mechanism may
be associated with enzymic reactions in the membrane, thus explaining the
action of certain of the inhibitors.  The enzymic steps are not considered to
represent separate rate-limiting factors.

WILBRANDT has derived a different set of equations which also fit the
kinetic and inhibition data.  His derivation is based on the assumption that
the carrier-glucose reactions are themselves enzymic in nature.  He assumes
that the inhibitors act directly on the carrier enzymes.

## Plant Cells

BROWN (1952) has reviewed studies on sugar uptake in segments of the
growing portion of roots of *Cucurbita*.  He cites experiments in which both
the rate of fructose uptake and respiration were measured as a function
of fructose concentration and $O_2$ tension.  The cellular concentrations of
reducing sugar were also measured.  The conclusion is reached that "sugar
absorption is an active process which requires the consumption of respiratory energy and which involves the operation of a transport mechanism
across a diffusion barrier." The data indicate that the diffusion barrier
must belocated at the outer surface of the protoplast rather than between
the cytoplasm and the vacuole.  During growth, although the surface is
increased four-fold, the rate of absorption remains relatively constant, thus
indicating that absorption occurs at specific sites constituting, at most,
25% of the surface.  A similar constancy despite increasing surface exists
for sucrose splitting, and for cell wall formation, both apparently mediated
by surface enzymes.  BROWN suggests that the surface sites in sugar absorption may also be enzymic in nature.  The evidence cited is compatible
with such a hypothesis but does not constitute proof.

## Aqueous Humor

Ross (1952) has studied the movement of glucose into the aqueous humor
from the blood and finds that it enters three times as rapidly as does urea,
much too repidly to be explained on a diffusion or lipid solubility basis.
In addition, insulin promotes the entry of glucose.  Ross suggests that

a phosphorylation by cell surface enzymes may be involved. However, HARRIS and GEHRSITZ (1949) found that there was no movement against a concentration gradient. DAVSON and DUKE-ELDER (1948), and HARRIS and GEHRSITZ, found that galactose, 3-methyl glucose, xylose, fructose, and arabinose penetrate rapidly but at the same rate. DAVSON and DUKE-ELDER suggest that some specialized mechanism must exist which allows the sugars to penetrate rapidly, but there is no direct evidence that it is enzymic in nature. The mechanism of insulin action on sugar is somewhat equivocal at the present time (see discussion on glucose uptake in muscle). Thus the specialized permeability of the aqueous barrier for sugars may or may not involve surface enzymes. The fact that many different sugars penetrate rapidly does not exclude an enzymic mechanism, because it has been shown that all of these sugars can be phoshorylated by intestinal homogenates (HELE 1950, CSAKY 1953).

### Intestine and Renal Tubule

The absorption of sugars in the intestine has been intensively studied, and the presence of an active transport mechanism well established. Detailed discussions can be found in VERZAR and McDOUGALD (1936) and in HÖBER (1946). For example, CORI (1925) and WILBRANDT and LASZT (1933) found that galactose and glucose are absorbed at a considerably higher rate than are other sugars. The latter authors found that with metabolic poisons the rates for glucose and galactose are the same as those for other sugars. In the case of glucose and galactose, the rate of absorption is constant over a range of sugar concentrations, whereas with the other sugars the rate of absorption is dependent on the sugar concentration. Such data have led to the conclusion that there is a passive diffusion of all of the sugars, but in addition in the case of glucose and galactose there is a metabolically linked active transport system. The existence of an active transport mechanism was unequivocally established by BARANY and SPERBER (1939) who showed that during absorption of glucose there was a movement from the lumen of the intestine to the blood against the concentration gradient. Furthermore, they showed that the absorption mechanism can be saturated by high substrate concentrations. Similar evidence has been reviewed by SMITH (1951) indicating the existence of an active transport of glucose in the renal tubule.

By what mechanism are the sugars reabsorbed? VERZAR and McDOUGALD (1936) and HÖBER (1946) developed the, thesis that resorption involves a phosphorylation mechanism. This conclusion is based primarily on studies of the effects of metabolic inhibitors such as iodoacetate and phlorizin, inhibitors known to affect phosphorylation reactions in both the intact intestine and in extracts of intestines. Other evidence has also been cited, such as the stimulating effect of phosphate on glucose resorption, and the presence of high concentrations of stainable phosphatases at the luminal border of the intestinal cells. DAVSON (1952), in discussing intestinal absorption, raises certain objections to the phosphorylation hypothesis. For example, Campbell and Davson (1948) found that 3-methyl glucose was actively transported by the intestine. However, this criticism does not eliminate the phosphorylation theory for Csaky (1953) recently showed that intestinal hexokinase can phosphorylate 3-methyl glucose. A more justified criticism of the inhibitor studies concerns the fact that metabolic inhibitors may inhibit active transport either by interfering with the energy supply required by the transport mechanism, or by interfering with the transport mechanism itself. Iodoacetate and phlorizin, although they may interfere with phosphorylation reactions, do not necessarily directly inhibit the transport mechanism itself, but may simply reduce the amount of energy available. For example, cyanide, which *does not* directly interfere with phosphorylation inhibits glucose uptake by the intestine (Kjerulf-Jensen and Lundsgaard 1940).

Are surface enzymes involved in the transport of sugars in the intestine and kidney? At the present time this question cannot be answered with any assurance. The movement of glucose from the lumen of the intestine to the blood is different from the movement of glucose into a single cell. In the intestine, two cell membranes plus the cytoplasm of the intestinal cell must be crossed. It is therefore difficult to divorce the role of the membranes from that of the cytoplasm. Various hypotheses have been presented in which the factors of permeability, diffusion, carriers, phosphorylation-dephosphorylation, and surface enzymes have been juggled in different combinations to provide an overall active transport across the intestinal cell. But the definitive experiments which would point out the correct combination has not yet been done. The various possibilities have been reviewed by Höber (1946), Rosenberg and Wilbrandt (1952), and Wilbrandt (1954). In regard to existence of surface enzymes, it must be decided whether the energy-requiring steps in the transport of glucose take place in the cytoplasm or in the periphery of the cell. If the former is the case, systems can be visualized in which sugar penetrates the membrane, perhaps aided by a non-enzymic carrier. Sugar could be phosphorylated in the cytoplasm at the luminal end of the cell, thus maintaining a concentration gradient across the luminal membrane. The glucose phosphate could then diffuse across the cell to be dephosphorylated at the other end of the cell, liberating free glucose which would diffuse across the membrane into the blood, perhaps aided by a carrier in the membrane. The basic difficulty with such a system is that the sugar concentration at the vascular end of the cell must always be higher than the blood sugar concentration in order for glucose to move toward the blood. As Wilbrandt (1954) points out, the tendency for back diffusion in the absence of membrane barriers would make such a system, at the least, very inefficient. It seems more likely therefore that all or part of the pumping mechanism which forces glucose to move against the concentration gradient is localized in one or in both membranes. Surface enzymes could participate either in phosphorylating-dephosphorylating systems, or in energy reactions coupled to a carrier system.

## Muscle

Glucose uptake by muscle has been the subject of intensive investigation for many years, with considerable emphasis on the role of insulin. By 1930 there were some 3000 references (CORI 1931). Probably as many papers have appeared since. Despite this tremendous output of research, it is still not certain exactly how sugars are taken up by the cell nor how insulin acts. However, there are certain indications that a transport of sugars occurs across the membrane of the muscle cell, which may involve enzymatic reactions, and that insulin influence the transport system. It is beyond the scope of the present paper to discuss the problem in any detail. A number of reviews concerning insulin and sugar uptake have appeared recently which cover the subject thoroughly (PARK 1952; STADIE 1951; KRAHL 1951; WICK, DRURY, and MACKAY 1951; STADIE 1954). Only a few of the recent, pertinent developments will be discussed here in any detail. It is only in recent years that it has been possible to sort out some of the factors in insulin action, that is, to separate the primary effects from secondary effects. There is some agreement now that insulin may have a single primary effect, namely, to increase the supply of sugar available for intracellular reactions. Such an increase can account for many of the effects of insulin on fat and protein metabolism (STADIE 1953; YOUNG 1953; STETTEN 1953; LUKENS 1953). The problem of insulin action thereby resolves itself into the question of how insulin can promote an increased supply of carbohydrate in the cell. Although there is no clear-cut answer to this question at the present time, considerable progress has been made.

The existing hypotheses concerning the action of insulin on sugar uptake can be divided into two groups. The first group includes those hypotheses which attempt to associate the insulin effect with activity of specific enzymes in the carbohydrate scheme. One enzyme that has frequently been implicated is hexokinase, the first enzyme to react with glucose. Enzymes associated with the generation of high energy phosphate have also received attention. The diverse effect of insulin could be explained if it increased the availability of ATP. In these hypotheses, it is generally assumed that glucose enters the cell by diffusion in response to a concentration gradient, and that it is then altered by intracellular enzymes. The removal of sugar by metabolism maintains the concentration gradient and the inward flow of glucose continues. A more recent hypothesis, first proposed by LEVINE, GOLDSTEIN, HUDDLESTUN, and KLEIN (1950) suggests that insulin acts on some transport system in the cell membrane which is responsible for the passage of sugars into the interior of the cell.

With regard to the events in sugar uptake which go on at the surface of the cell, three possibilities exist:

(1) The sugar may pass through the membrane by diffusion in response to its concentration gradient.

(2) The sugar may be carried across the membrane in the form of a complex with a membrane constituent. If the driving force is the con-

centration gradient, then the transport system may be non-enzymatic in nature.

(3) The sugar may be actively transported across the membrane with metabolically derived energy as the driving force. The transport mechanism may involve direct enzymatic alteration of the sugar in the membrane, or it may involve a carrier system coupled to enzyme reactions.

Many attempts have been made to associate the action of insulin with specific enzymes in glucose metabolism (see review of Krahl 1951). The

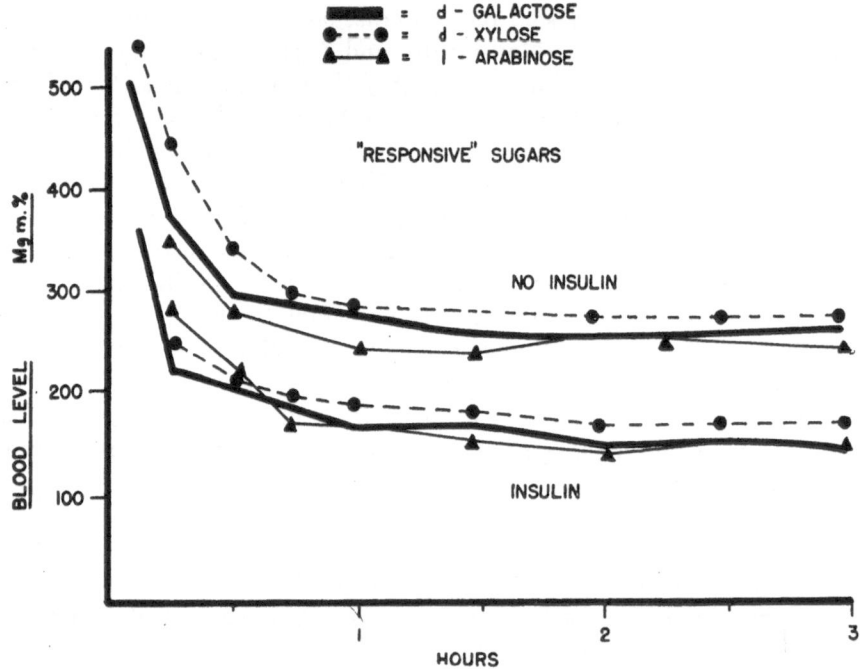

Fig. 17.   Effect of insulin on distribution of sugars in eviscerated-nephrectomized dogs.
Data previously reported for the effect of insulin on distribution of d-galactose (1) are represented by heavy lines. Note that d-xylose and l-arabinose exhibit identical behavior. For all 3 sugars an equilibrium is established, in the absence of insulin in a body space of about 40—45 %. The presence of insulin promotes a distribution of these 'responsive' sugars throughout 65—70 % of body weight, i. e., total body water. All curves are the averages of 4—6 animals.
(Goldstein et al. 1953; courtesy Amer. J. Physiol.)

results are inconclusive at best. For example, Cori (1945–46) found that a certain percentage of extracts from animals previously inhibited by adrenal and pituitary factors, could be reactivated by insulin. Others have obtained variable results or negative results. The results with tissue homogenates and extracts have been so variable or inconclusive that Stadie (1954) has suggested that "retention of cellular integrity is necessary for hormonal action upon enzyme systems," and that the hormone must combine with a receptor in the cellular structure.

Studies of insulin effects in intact tissues reveal that many changes occur. Glucose is oxidized more rapidly (Villee, White, and Hastings 1952). Phosphate turnover is increased (Sacks 1952; Sacks and Sinex 1953).

Glycogen synthesis is increased (GEMMILL and HAMMAN 1941). The question remains, however, as to cause and effect. Are these effects due to the influence of insulin on an enzyme or are they due to an increased intracellular supply of sugar? Some light is thrown on the situation by studies of the specificity of insulin. CORI and CORI (1929), in early studies using eviscerated rats, found that glucose uptake was enhanced by insulin, but fructose and mannose uptakes were not. LUNDSGAARD (1939) similarly found that insulin influenced glucose uptake but not fructose uptake in the perfused hind limb of the cat. Both observations were consistent with

Table 7. *The Specificity of the Insulin Effect on Sugar Uptake.*

|  | Eviscerated Rat[1] | Hind Limb Cat[2] | Eviscerated Dog[3] | Eviscerated Rabbit[4] | Rat Diaphragm[5] | Rat Diaphragm[6] | Rat Diaphragm[7] |
|---|---|---|---|---|---|---|---|
| d glucose ..... | + | + | + | + | + | + | + |
| d fructose..... | — | — | — | — | + | + | + |
| d mannose .... | — |  | — | + |  |  |  |
| d galactose ... |  |  | + | + | + | + |  |
| l sorbose ..... |  |  | — |  |  |  |  |
| d arabinose ... |  |  | — |  |  | — |  |
| l rhamnose ... |  |  | — |  |  |  |  |
| d xylose ...... |  |  | + |  |  | ± |  |
| l arabinose ... |  |  | + |  |  | ± |  |
| d sorbitol ..... |  |  | — | — |  |  |  |
| gluconic acid .. |  |  |  | — |  |  |  |

[1] CORI and CORI 1929.
[2] LUNDSGAARD 1939.
[3] GOLDSTEIN, HUDDLESTUN, and LEVINE 1953.
[4] WICK and DRURY 1953 a.
[5] MACKLER and GUEST 1953.
[6] HAFT and MIRSKY 1953.
[7] DEMIS and ROTHSTEIN 1954.

the hexokinase site of insulin action. Recently, however, evidence has accumulated that other sugars are also influenced by insulin. LEVINE, GOLDSTEIN, HUDDLESTUN, and KLEIN (1950) found that galactose uptake is markedly enhanced in eviscerated animals by insulin despite the fact that galactose is not utilized by such preparations. Later, GOLDSTEIN, HENRY, HUDDLESTUN, and LEVINE (1953) found that two non-utilizable pentoses, 1-arabinose and d-xylose exhibited the same behavior as galactose (Fig. 17). They postulated therefore that insulin could not be acting on an enzyme because enzymes to metabolize galactose, arabinose and xylose are not present. These results and conclusions were essentially confirmed by WICK and DRURY (1953 b), using radioactive galactose in eviscerated rabbits. The same authors (1953 a) found that mannose uptake was also influenced by insulin, although this sugar is not appreciably oxidized.

The specificity pattern for insulin action on various preparations is compiled in Table 7. There are certain discrepancies in the data. In the eviscerated dogs and rats, d-mannose is not affected by insulin, but in the eviscerated rabbit it is. In the eviscerated dogs, rabbits, rats and cats, d-fructose is not affected; in isolated diaphragms it is. Perhaps species differences in the response to insulin may account for the discrepancies. However, in the case of fructose, the absence of an insulin effect in the eviscerated preparations may be due to the fact that glucose was present in the perfusate. Mackler and Guest (1953) have shown in isolated rat diaphragm that glucose represses the insulin effect on fructose.

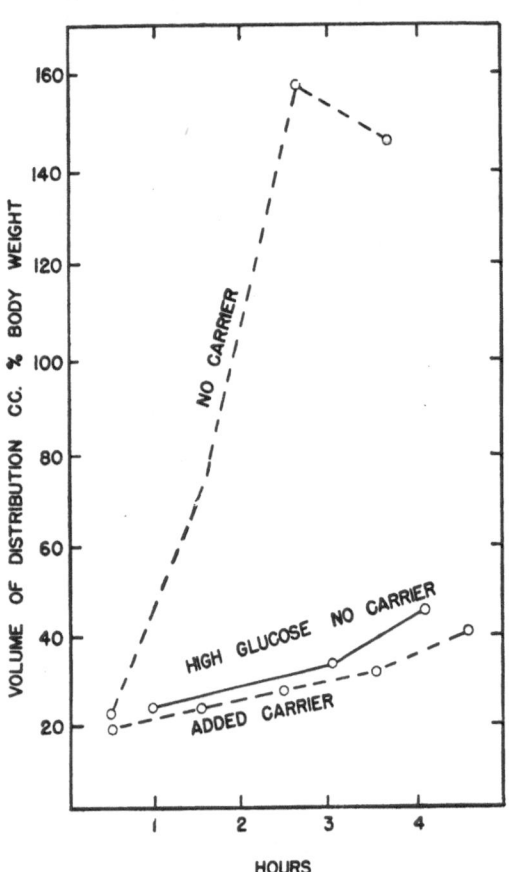

Fig. 18. Effect of carrier galactose (1 gm./kg.) and of a high plasma glucose concentration (800 mg./%) on volume of distribution of galactose-1-C[14] in insulin-treated animals.

(Wick and Drury 1953; courtesy Amer. J. Physiol.)

Regardless of species differences, it is obvious that no single kinase is involved for all sugars influenced by insulin. Hexokinase of muscle can only phosphorylate glucose and fructose, but not the other sugars tested. In fact, the oxidation of some of the sugars by any pathway is perhaps too low to account for the insulin effect on an enzyme. For galactose, it is less than 10% of that for glucose (Wick and Drury 1953 b). However, it should be pointed out that enzymic transformation of galactose may lead to glycogen formation rather than oxidation. It has been found, for example, in rat diaphragm that the same proportion of galactose and fructose taken up is converted to glycogen, as is the case for glucose (Demis 1953). In view of the fact that all of the work with the "non-metabolized" sugars has been measured in terms of their disappearance from the blood or medium, it seems a little premature to assume that all of the sugar which disappears will be found unaltered in the interior of the cell. In any case, the insulin specificity is incompatible with any known, simple, pattern of enzyme activity. Furthermore, Wick and Drury (1953 b) have shown that glucose and galactose compete for the insulin-stimulated mechanism (Fig. 18). Yet no presently known enzyme can react with both sugars.

Does insulin act on the cell surface? Recent experiments throw some light on this question. Drury and Wick (1951), using radioactive glucose in the eviscerated rabbit, found that the distribution of freely diffusible glucose was in the extracellular spaces only. They inferred that only small amounts of free glucose could be present in the intracellular spaces. PARK and JOHNSON (1953), using eviscerated rats, analyzed tissues for glucose by a specific enzymic method. With a wide range of serum glucose levels (30 to 1000 mg.) they found that in the absence of insulin there was no free glucose in the cells. That found in the tissues was exclusively extracellular. These results indicate that as rapidly as glucose passes through the membrane into the cell it is metabolized. The enzyme capacity within the cell is adequate to handle all the sugar that enters, even with high external glucose concentrations. Therefore, the movement of glucose through the membrane rather than the internal enzyme capacity must constitute the rate limiting step. Insulin could not be acting on the internal enzymes to speed up the glucose uptake because these enzymes are already adequate to handle all the glucose that reaches them. Insulin can only stimulate sugar uptake by acting to speed up its movement through the membrane. PARK and JOHNSON found that when insulin is added, some free glucose is found in the heart and diaphragm, but not in striated muscle of carcass. Thus insulin can apparently speed up the transport through the surface to the extent that the enzyme complement of the cell no longer handle all of the sugar presented to it. The hypothesis that insulin acts on a cell surface mechanism is also compatible with the studies of STADIE (1954) of insulin binding by the cells. STADIE finds that there are a specific limited number of sites with which insulin can combine. It seems reasonable that these sites are on the surface of the cell rather than in the interior, otherwise higher insulin concentrations should result in a greater uptake, with no saturation. The rapidity with which insulin is fixed by muscle also supports the concept of a surface binding. STADIE found that a few seconds of exposure was sufficient to fix appreciable quantities of the hormone. It is not likely that in this short time insulin could diffuse into the interior of the cell through a membrane which is impermeable to molecules of much smaller size. There is no evidence that proteins in general can pass through the membrane of muscle cells. To the contrary, MILLER (1954) has shown, by labeling the cellular proteins of muscle with $C^{14}$ amino acids, that proteins do not pass through the cell membrane at all (see discussion on the synthesis of plasma proteins). Unless one postulates an unique mechanism for transporting insulin into the cell, it must be assumed that insulin combines on the cell surface.

Assuming then, that insulin acts on a sugar transport system in the cell surface, what can be said concerning the properties of the system? It has already been pointed out that a certain pattern of specificity for sugars is involved. Thus, the transport must involve specific substances which can combine readily with some sugars and not with others. This excludes any non-specific mechanism such as diffusion or lipid solubility, or non-specific adsorption. Furthermore, WICK and DRURY (1953 b) showed that in the presence of insulin, with low galactose concentrations, a much greater percentage of the sugar was taken up by muscle than with high concentrations (Fig. 18), indicating that the transport mechanism had a limited carrying capacity which could be saturated by high concentrations of sugar. This points to a limited capacity of the transport mechanism.

There is some evidence that the surface transport system is an active transport driven by cell metabolism rather than by the diffusion gradient. For example, Wick and Drury found in eviscerated rabbits that low concentrations of galactose were apparently moved against the concentration gradient, achieving a volume of distribution of 150% of cell water (Fig. 18). A similar finding was reported for rat diaphragm (Demis 1953). In each case it was assumed that the galactose, which disappeared from the extracellular environment, was present in the interior of the cell in the form of free galactose. This may not be the case for Demis (1953) has found that in the isolated rat diaphragm, galactose is partially converted to glycogen.

Other evidence of a coupling of the insulin effect to metabolism is found in the studies of Demis and Rothstein (1954). They confirmed the earlier studies of Walaas and Walaas (1952) that glucose is taken up by rat diaphragm just as rapidly under anaerobic as under aerobic conditions, i. e., anaerobiosis does not interfere with the normal uptake of glucose. However, the insulin effect only takes place in the presence of oxygen and not under anaerobic conditions. The absence of an anaerobic insulin effect is not due to irreversible damage to the muscle associated with anaerobiosis, for the same muscles which fail to show an insulin effect anaerobically will do so if returned to air. It was concluded that insulin does not act directly on the normal mechanism for glucose uptake, but acts through, or in association with, some cellular factor dependent on aerobic metabolism. There is other evidence that the insulin pathway is not identical with the normal pathway of sugar uptake. Mackler and Guest (1953) found that glucose had no effect on the uptake of fructose in the absence of insulin, but that glucose inhibited the increment of fructose uptake associated with insulin. Haft and Mirsky (1952) found that iodoacetate in concentrations which did not inhibit glucose uptake, did inhibit the insulin effect.

The events at the cell surface in glucose uptake without insulin have been characterized using mercury as an inhibitor. With low concentrations of mercuric chloride, the metal ion is rapidly bound by the surface of the muscle cell. With higher concentrations, it slowly penetrates into the interior. Associated with the surface binding of mercury, there is a complete block of glucose uptake, but no effect on respiration. Associated with the penetration into the interior of the cell, there is an inhibition of respiration. Because the surface phenomena in sugar uptake are inhibitable by small amounts of mercury, it is suggested that an enzyme reaction is involved (Demis 1953).

## Transport of Disaccharides

Dormer and Street (1949) and Street and Lowe (1950) found that excised tomato roots could utilize sucrose at a much greater rate than either glucose, fructose or an equimolar mixture of the two. The utilization of sucrose is associated with the appearance of glucose and fructose in the medium, the amount of fructose corresponding to the theoretical yield expected from the breakdown of sucrose, but the amount of glucose considerably less than theoretical. Thus, only the glucose moiety of sucrose is taken up. This, coupled with the fact that glucose by itself is only taken up slowly, indicates that some special mechanism for simultaneously taking up the glucose moiety of sucrose, but leaving fructose behind in the medium. DORMER and STREET postulate the phosphorolysis of the sucrose mediated by an enzyme in the cell surface which results in the production of glucose phosphate plus fructose. The former is able to pass into the cell whereas the fructose remains behind in the medium. DORMER and STREET eliminate the possibility that a surface invertase activity precedes the uptake of the glucose moiety. They suggest that a surface invertase may be present in addition to the phosphorylase, thus explaining the appearance of a certain quantity of glucose in the medium. Additional evidence is presented which supports the concept of an enzyme-mediated transport of sugar. For example, there is a well defined pH optimum at 4.8. The addition of phosphate increases the sucrose absorption by phosphate deficient roots. Phlorizin, an inhibitor of phosphorylating reactions, inhibits sucrose absorption competitively. That is, the inhibition is partly reversed by higher sucrose concentration.

DOUDOROFF (1951) has studied the uptake of disaccharides by a number of microorganisms. An enzyme capable of catalyzing the phosphorylytic breakdown of sucrose has been isolated from *Pseudomanas saccharophila* and *P. putrefaciens*. The products are glucose-1-phosphate and fructose. When sucrose is fed to intact cells this substrate is rapidly utilized in its entirety. Yet, neither free glucose nor free fructose can be utilized at comparable rates. Thus the cell possesses a specialized ability to take up fructose and glucose combined as sucrose. Furthermore, sucrose, labeled with $C^{14}$ in the fructose moiety, does not exchange with unlabeled fructose given at the same time.

If the cells are given glucosido-sorbose rather than sucrose, it is utilized as rapidly as sucrose. However, only the glucose portion of the molecule is used; the sorbose is quantitatively recovered in the medium. The cell has a high degree of discrimination. It will rapidly take up bound glucose or fructose, but not free glucose or fructose or bound sorbose.

DOUDOROFF discusses the possibility that the specific "permeability of the cells to certain disaccharides may be due to the fact that specific enzymes such as sucrose phosphorylase may act as carriers to transport the sugars across the cell membrane."

## Transport of Phosphate

The role of phosphate in metabolism has been studied intensively, particularly in relation to the phosphorylation reactions in carbohydrate metabolism and the utilization of phosphate bond energy in bio-synthetic reactions. When phosphate metabolism is studied in intact cells, dramatic changes in the total content of phosphate, or in the content of any particular phosphate compound do not necessarily occur, even though a phosphate cycle is operative. However, if $P^{32}$ labeled orthophosphate is added to cells, the various phosphate fractions become labeled with the isotope, indicating a high rate of turnover of the components of the cycle. Much attention has been directed toward the identification of the reactions of the phosphate cycle itself. Less attention has been directed toward the mechanism by which phosphate moves from the medium across the cell membrane into the interior of the cell. Many investigators have simply assumed that the phosphate diffuses across the membrane and that it participates in the phosphate cycle once it has reached the interior of the cell. There is, however, a growing body of evidence which indicates that the simple diffusion concept is an oversimplification. In fact, in many cells it can be shown that the membrane is essentially impermeable to phosphate and that phosphate uptake only occurs during the metabolism of exogenous substrates. In a few cases, surface enzymes are definitely implicated.

Before discussing phosphate uptake by any particular cell, it would be well to define the term. *Phosphate uptake* is here defined as the movement of inorganic orthophosphate from the medium into the cell with a consequent increase in the total phosphate content of the cell. In order to demonstrate a *phosphate uptake,* chemical determinations of the phosphate content of the cell or of the medium must be made. When isotope techniques are used without concomitant chemical measurements of the phosphate uptake, it is not possible to determine how much of the $P^{32}$ that gets into the cell is associated with a phosphate uptake and how much with an exchange of cellular phosphate with $P^{32}$ from the medium (hereafter called *phosphate exchange*). In many of the studies, using $P^{32}$ labeled phosphate, the appearance of $P^{32}$ in the cell has been called "phosphate uptake" even though it is not clear whether the phenomenon represents an uptake or an exchange. To avoid confusion in the present discussion, any measurements of the appearance of $P^{32}$ in the cell without a concomitant measurement of chemical phosphate will be called $P^{32}$ *transfer.*

## Microorganisms

Kamen and Spiegelman (1948), in their review of phosphate uptake by microorganisms, cite considerable evidence linking phosphate uptake and, or, exchange, with the metabolism of the cell. They point out many of the difficulties inherent in studying phosphate uptake by isotope techniques, difficulties associated with non-heterogeneity of the various phosphate fractions and with the hydrolysis of labile compounds. Nevertheless, they conclude that the available data are consistent with "the assumption that the primary mechanism of the entrance of phosphate is via an esterifying mechanism." No suggestions are made concerning the location of the esterification enzymes in the cell.

Studies on permeability to phosphate are inconclusive, with considerable variations encountered in different organisms. Certain organisms seem to be permeable to phosphates. KAMEN and SPIEGELMAN (1948) quote examples of phosphate leakage, during washing, from cells such as algae, fungi and bacteria. Certain bacteria and protozoa can take up phosphorylated compounds and utilize them as substrates (ROBERTS and WOLFFE 1951). In contrast, MITCHELL (1953) has shown by volume of distribution studies that the membrane of *M. pyogenes* is impermeable to inorganic phosphate. However, if $P^{32}$ labeled phosphate is used, a rapid exchange of phosphate is observed. In the case of the yeast cell, the membrane is not only impermeable to phosphate, but rate of inward and outward exchange is exceedingly low (HEVESY, LINDERSTRÖM-LANG and NIELSEN 1937; MULLINS 1942). In view of the variations in permeability to phosphate cited above, it is unwise to generalize and best to discuss the properties of phosphate uptake by specific organisms.

HOTCHKISS (1943) studied the uptake of phosphate by *Staphylococcus* and found that metabolism is intimately concerned. With no substrate present, phosphate was not taken up from the medium even with a concentration of phosphate as high as 0.3 M. But if glucose were added, considerable phosphate was absorbed. In some cases, all of the measureable phosphate was removed from the medium indicating that the inward movement is against the concentration gradient, for bacterial cells contain measurable concentrations of inorganic phosphate. Substrates other than glucose, such as glycerol, lactate, alcohol and pyruvate, did not invoke a phosphate uptake even though they were metabolized. Thus reactions specific to glucose metabolism are necessary. It seems evident that the staphylococcal cell possesses an active transport mechanism for absorbing phosphate against the concentration gradient, which can only be energized by sugar metabolism. Furthermore, the transport system is located at least in part on the surface of the cell, for phosphate uptake is inhibited by gramicidin, which, according to HOTCHKISS, acts on the cell surface.

Phosphate uptake has been most intensively studied in the yeast cell. As already pointed out, the yeast cell membrane is impermeable to phosphates. Cells suspended in $P^{32}$ labeled inorganic orthophosphate take up the isotope only very slowly (HEVESY, LINDERSTRÖM-LANG and NIELSEN 1937, MULLINS 1942). Nor can they take up organic phosphate (ROTHSTEIN and MEIER 1948, 1949). Movement of phosphate from the cell to the medium is also exceedingly slow (HEVESY and ZEHRAN 1946; JUNI, KAMEN, REINER, and SPIEGELMAN 1948).

Only when exogenous substrate is present is there a rapid uptake of phosphate, as measured either by isotope techniques or by chemical analysis. The amount of phosphate taken up is dependent not only on the presence of substrate, but also on the phosphate content of the cell (JEENER and BRACHET 1944) and on the presence of $K^+$ and $Mg^+$ (SCHMIDT, HECHT, and THANHAUSER 1949). The rate of uptake is markedly dependent on temperature (HEVESY, LINDERSTRÖM-LANG, and NIELSEN 1937; MULLINS 1942). It is markedly enhanced by riboflavin (NICKERSON and MULLINS 1948). The association of phosphate uptake with metabolic reactions is also evident from inhibitor studies.  KAMEN and SPIEGELMAN (1948) point out that substances that have been implicated as inhibitors of reactions involving esterification of phosphate, such as iodoacetate, arsenate, azide and dinitrophenol, can prevent phosphate uptake.  SPIEGELMAN, KAMEN, and SUSSMAN (1948) have investigated the effects of azide in some detail.  Concentrations of this agent which do not influence the rate of sugar fermentation, not only block the exchange of extracellular and intracellular phosphate, but also suppress the esterification of cellular phosphate.  Esterification as a basic factor has been implicated both by the intimate association of phosphate uptake with metabolism, and by the action of inhibitors.

Other properties of phosphate uptake by yeast have been described by GOODMAN and ROTHSTEIN (1954).  They found that the fermentable sugars, glucose, fructose and mannose will all induce a phosphate uptake, under either aerobic or anaerobic conditions.  In contrast, other substrates such as lactate, acetate, pyruvate and alcohol have little effect, even though these compounds are respired just as rapidly as are the sugars.  Thus the phosphate uptake is coupled specifically with glycolytic reactions rather than oxidative phosphorylation reactions of the Krebs cycle, and its associated aerobic reactions.  Although phosphate uptake occurs only when sugar is taken up and metabolized, the reverse is not true.  Glucose uptake proceeds in the absence of extracellular phosphate but the rate is reduced to about 80% of that in the presence of phosphate.  Thus phosphate influences sugar uptake but is not a requirement.  There is no simple stoichiometric relationship between the rate of glucose uptake and the rate of phosphate uptake.  Under optimal conditions of potassium concentrations and pH, $^1/_{10}$ mol of phosphate is taken up for every mol of glucose.  Many factors will alter this ratio.

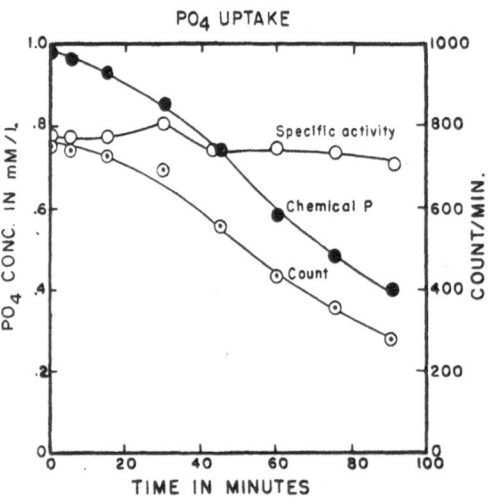

Fig. 19.   The uptake of phosphate by cells actively metabolizing glucose.

The yeast concentration was 30mg./ml. and the pH 4.5. Glucose was added at zero time to make an initial concentration of 0.16 M/l.

ROTHSTEIN and MEIER (1949) studied phosphate uptake by measuring the disappearance of phosphate by chemical determinations as well as by measurements of $P^{32}$ activity.  The percentage of phosphate that disappeared from the medium in the presence of glucose was the same when measured chemically as when measured by $P^{32}$ activity, indicating no change in the specific activity of residual phosphate (Fig. 19).  It is evident, therefore, that the extracellular phosphate is not diluted by exchange for cellular phosphate, even in the presence of glucose when a rapid uptake of phosphate occurs. In other words, phosphate goes into the cell, but none comes out. It is essentially an irreversible process.

The dynamics of phosphate uptake provide further information. As the extracellular phosphate concentration is raised, the rate of uptake increases asymptotically (Fig. 20). Such behavior indicates that some component of the phosphate uptake mechanism is present in a limited quantity. It becomes saturated and limits the overall rate as the phosphate concentration is raised. The existence of a saturation phenomenon indicates that phosphate must combine with a cellular component during its transport.

During the course of phosphate uptake the concentration of inorganic orthophosphate within the cell remains relatively constant at a level of .01 to .02 M. The increased phosphate content of the cell is represented

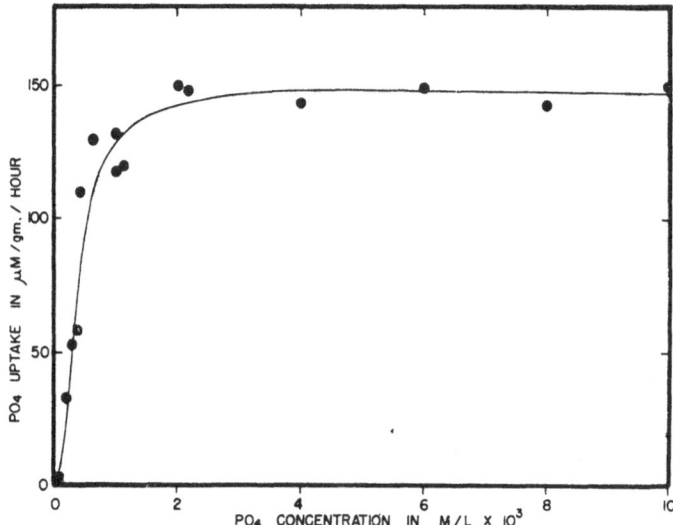

Fig. 20.   The effect of phosphate concentration on the rate of phosphate uptake by yeast.

primarily by an increase in the amount of metaphosphate (SCHMIDT 1951). Despite the relatively high intracellular orthophosphate concentration, there is a rapid rate of phosphate uptake when the extracellular concentration is as low as $1 \times 10^{-4}$ M (Fig. 20). Thus phosphate is taken up against a concentration gradient as high as 100 to 1. These data leave no question that the uptake of phosphate involves an active transport system, a conclusion supported by a high temperature coefficient, by the coupling to metabolism, and by the irreversible nature of the process.

By what mechanism does the phosphate pass through the cell membrane? The yeast membrane is impermeable to phosphate to such an extent that no exchange of phosphate occurs even when phosphate is taken up. There is no direct communication between the inorganic orthophosphate of the medium and that inside the cell. However, when sugar is present, phosphate is forced into the cell against the concentration gradient. It seems highly likely, therefore, that some carrier system must be present in the periphery of the cell which combines with the phosphate and transports it into the cell. How is the carrier system energized? It is conceivable that a non-enzymic combination of phosphate with a carrier substance in the cell membrane could occur. Perhaps the metabolic reactions simply destroy or synthesize the carrier substance at opposite sides of the membrane, thus achieving transport. However, in the case of phosphate, it seems more likely that the carrier substance is itself a phosphate compound produced by enzymatic esterification reactions associated with sugar metabolism. It is significant in this regard that sugars are the only substrates that invoke a rapid phosphate uptake and that fermentation is as effective as respiration. It has been shown in the previous discussion on sugar uptake that the fermentive enzymes are located in the periphery of the cell and are thus available at the surface to esterify phosphate. The reaction which must be implicated is the one catalyzed by phosphoglyceraldehyde dehydrogenase. This can be written

3-phosphoglyceraldehyde $+ PO_4 = 1, 3$-diphosphoglycerate,

1, 3-diphosphoglycerate $+ ADP = 3$-phosphoglycerate $+ ATP$.

Thus the inorganic phosphate would be transferred rapidly to ATP from whence it would pass into the various metabolic channels as found by Juni, Kamen, Reiner, and Spiegelman (1948). The actions of azide and dinitrophenol are also consistent with the proposed mechanism. These inhibitors prevent phosphate uptake without inhibiting the fermentation of glucose. They also prevent the esterification of phosphate by the glyceraldehyde dehydrogenase reaction, without inhibiting the oxidative reaction (Spiegelman, Kamen, and Sussman 1948). It is of interest to note that in red blood cells, 2, 3-diphosphoglycerate (a compound in equilibrium with 1, 3-diphosphoglyceraldehyde) and ATP seem to serve as precursors for cellular inorganic orthophosphate (see next section).

### Red Blood Cells

The $P^{32}$-transfer by red blood cells is markedly dependent on temperature (Eisenman, Ott, Smith, and Winkler 1940; Gourley and Gemmill 1950). It requires the presence of glucose but is not influenced by the presence or absence of oxygen. It is depressed by iodoacetate or NaF, but not by DNP or azide (Gourley 1952 b). On the basis of such evidence it has been suggested that phosphorylating reactions in metabolism are associated with the uptake of inorganic phosphate, and attempts have been made to identify the specific reactions involved. Fractionations were made of various phosphate compounds of the human red cell at various times after the addition of $P^{32}$ (Gourley 1952 a). It was found that for a period of several hours the specific activity of the labile P of ATP was considerably higher than that of the cellular inorganic phosphate, but after 8 hours they were equal. It seems likely, therefore, that the outside inorganic phosphate does not equilibrate directly with the cellular inorganic phosphate, but that ATP must first be formed and that this compound acts as a precursor for the inorganic phosphate of the cell. Support for this concept was also obtained by inhibitor studies (GOURLEY 1952 b). Iodoacetate and fluoride reduce the inflow of phosphate into the cell and also reduce both the quantity of ATP and its rate of turnover. It was concluded that phosphorylation reactions in the membrane of the cell involving ATP synthesis must first proceed, followed by release of inorganic phosphate into the cell by hydrolysis of ATP or other organic compounds.

GOURLEY and MATSCHINER (1953) found that not all of the $P^{32}$ was taken up by the mechanism described above. An additional system with a different time constant was also present, involving simple diffusion of phosphate into the cell.

GOURLEY, in determining specific activities of cellular compounds, used a method developed by SACKS (1949) for liver and applied it to red blood cells. Precipitation and hydrolysis techniques for separating phosphate compounds of tissues, although useful, are notoriously difficult to quantitate because of such problems as incomplete precipitations, co-precipitations and incomplete hydrolyses. Recently the uptake of phosphate by the red cell has been studied using paper chromatographic techniques to separate the phosphate compounds (PRANKERD and ALTMAN 1954). GOURLEY's finding that ATP acts as precursor for the cellular inorganic phosphate has been confirmed, but in contrast

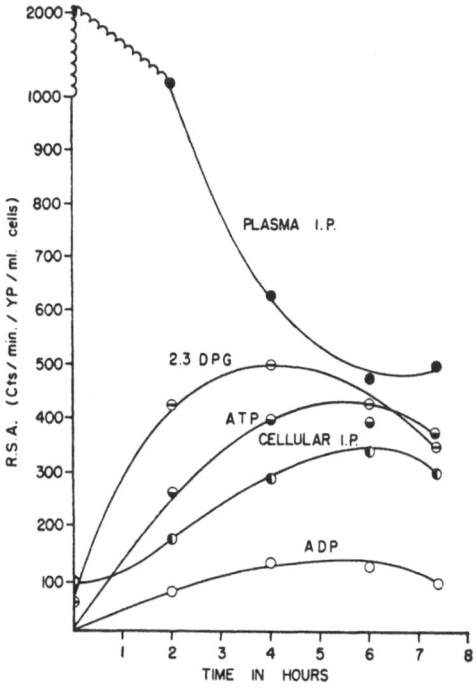

Fig. 21.   The uptake of $P^{32}$ into various compounds in the red blood cell.

Plasma IP (plasma inorganic phosphate), 2.3 DPG (2, 3-diphosphoglycerate), ATP (adenosine triphosphate), ADP (adenosine diphosphate), and cellular IP (cellular inorganic phosphate).

(PRANKERD and ALTMAN 1954; courtesy of the authors.)

to GOURLEY's results, the 2, 3-diphosphoglycerate seems to be a precursor to ATP, reaching a higher specific activity earlier (Fig. 21). This finding seems entirely reasonable, because ATP cannot directly take up orthophosphate. Red blood cells will take up phosphate under anaerobic conditions in the presence of glucose. Some reaction associated with glycolysis must be responsible. The only glycolytic reaction which esterifies phosphate is the reaction of 3-phosphoglyceraldehyde with phosphate and transfer of phosphate to ATP by the glyceraldehyde dehydrogenase reaction. However, the intermediate formed, 1, 3-diphosphoglyceraldehyde is unstable. RAPOPORT and LUEBERING (1950) have shown that this compound is in equilibrium with 2, 3-diphosphoglycerate which is very stable. On this basis, the precursor relationship of 2, 3-diphosphoglycerate, ATP and cellular inorganic phosphate are compatible with the concept of a pickup of extracellular phosphate by the 3-phosphoglycerate and gradual release to ATP, with eventual equilibration with the inorganic phosphate in the interior of the cell. Further support for the concept is given by Altman (1954), who found that the 2, 3-diphosphoglycerate is located in large part in the stroma fraction of the red cell, which presumably includes or constitutes the cell membrane.

## Sea Urchin Eggs

Lindberg (1950) investigated the $P^{32}$-transfer by unfertilized and fertilized sea urchin eggs (*Paracentrotus boidus*). He measured the rates of uptake, and the specific activities of the cellular ATP and orthophosphate. In unfertilized eggs, the specific activity of the ATP rose more rapidly than that of the orthophosphate, but after 30 minutes both fractions reached the same plateau which was maintained for the duration of the experiment. However, the specific activity for the cellular phosphates, after equilibrium had been reached, was considerably lower than that of the extracellular phosphate. Apparently, in this case, the extracellular phosphate is picked up by surface ATP. It then equilibrates with surface orthophosphate but does not mix with the phosphate of the general cytoplasm. In fertilized eggs, much larger amounts of phosphate are taken up by ATP and penetration into the interior of the cell occurs.

## Animal Tissues

There has been considerable controversy concerning the mode of uptake of phosphate. Most of the studies have been performed by measuring $P^{32}$-transfer and its incorporation into various organic fractions of the tissue. On the basis of such experiments, Sacks (1953) has suggested that orthophosphate is taken up by incorporation into organic compounds in the cell membrane with the orthophosphate inside the cell being derived from these compounds. Others have opposed this view and suggested a diffusion mechanism (Furchgott and Shorr 1943; Kalckar, Dehlinger, and Mehler (1944). The controversy will not be discussed here in any detail (see Greenberg 1952). It is based on differences of opinion in evaluating and interpreting the available data. For example, tissues contain considerable interstitial space. When phosphate is added to the blood or to an isolated muscle, it first equilibrates with the interstitial space, which in turn equilibrates with the cells. The inorganic phosphate content and radioactivity (in experiments with $P^{32}$) must be obtained either by calculating a correction for the interstitial phosphate, or by washing it out. Neither approach is completely satisfactory (Ennor and Rosenberg 1954 a and 1954 b). The precursor relationships between the various phosphate compounds of muscle and liver are not altogether clear for another reason. There is ample evidence that cells are compartmentalized. Any given phosphate compound extracted from the cell may have come from a number of compartments. If only one of the compartments participates in phosphate turnover, then the phosphate compounds from that compartment are diluted by the same compound extracted from other inactive compartments. Thus the turnover rates for a given compound, as well as the precursor relationships are distorted or obscured. The problem of non-homogeneity of phosphate compounds is particularly apparent in the case of kidney, studied by DRATZ and HANDLER (1952). Taken as a whole, the $P^{32}$-precursor data for liver, muscle, and kidney do not either conclusively prove or disprove the thesis that the sequence of events is:

> extracellular orthophosphate → organic phosphates in the
> cell membrane → intracellular orthophosphate.

Some of the data are certainly consistent with the hypothesis. None of the data disprove the hypothesis.

Support for the membrane-phosphorylation concept is found in other studies. POPJAK (1950) perfused rabbit livers with plasma containing $P^{32}$ labeled inorganic phosphate. Both the plasma and liver were analyzed for phosphate. Despite the fact the concentration of phosphate in the liver cells was four times as high in the plasma, considerable phosphate was taken up. Thus the phosphate is actively transported against the concentration gradient. Furthermore, the uptake was inhibited by azide, which is known to inhibit oxidative esterification of phosphate. POPJAK concluded that phosphate must be taken up by metabolic reactions involving phosphorylation at the cell membrane.

In nearly all of the recent studies of phosphate uptake by muscle, measurements have been made in terms of $P^{32}$ transfer rather than phosphate uptake. Before the availability of isotopes, EGGLESTON (1933) studied the equilibration of frog muscle with various phosphate environments. A constant fraction of the muscle was found to equilibrate, which was about 20%, the volume now attributed to interstitial space. Recently, HARRIS (1953) has studied the equilibration of frog muscle with phosphate, using $P^{32}$. In the absence of substrate, the membrane of the muscle cells behaves as though it were impermeable to phosphate. Not only did the muscle fail to leak or take up phosphate, but there was no exchange of $P^{32}$ labeled phosphate for cellular phosphate. Nevertheless, in the presence of glucose, many studies have shown that $P^{32}$ is transfered into the cell and incorporated into cellular compounds. The passage of phosphate through the membrane is metabolically linked. Glucose seems to be a requirement.

The strongest support for the membrane phosphorylation mechanism is given by CAUSEY and HARRIS (1951). Frog muscle was exposed to $P^{32}$ labeled orthophosphate, washed free of adhering solution. Analysis of phosphate fractions was made after various pretreatments and washing procedures. In addition, radio-autographs were prepared. The pictures conclusively show that the phosphate is localized in the periphery of the cell. Furthermore, this peripheral phosphate is present as organic phosphates which were not free to diffuse into or out of the cell. Thus the phosphate uptake by muscle must involve phosphorylation reactions at the periphery of the cell. The specific enzymes involved in the cell surface phosphorylation would seem to be those associated with sugar metabolism.

## Relationship of Phosphate Transport and Sugar Transport

In the preceding sections, evidence has been presented that surface enzyme reactions are involved in the uptake of sugar by yeast. In the case of mammalian tissues such as red cells and muscle, there is evidence that a surface transport mechanism is present, but the direct participation of enzymes in this transport is a matter of speculation. Independent evidence has been cited to support the conclusion that phosphate uptake or $P^{32}$-transfer by yeast, bacteria, sea urchin eggs, human red blood cells and muscle, involves the formation of phosphorylated compounds on the surface of the cell by enzymic reactions. The question naturally arises as to the connections, if any, between the surface transport mechanisms for sugars and those for phosphates.

Many examples can be given of inter-relationships between sugar uptake and phosphate uptake or $P^{32}$-transfer. In some microorganisms (*Staphylococcus* [Hotchkiss 1943] and yeast [Goodman and Rothstein 1954]), phosphate uptake occurs only when substrate is present, and furthermore this substrate must be a fermentable sugar. In human red blood cells (Gourley 1952) and in muscle, $P^{32}$-accumulation is dependent on the presence of glucose but other substrates have not been tested. Phosphate transfer occurs in the presence of glucose under either aerobic or anaerobic conditions in both red cells and yeast. In muscle, insulin, which increases the uptake of glucose, markedly accelerates the turnover of various phosphate fractions, although it does not alter the actual amount of these fractions (Sacks and Sinex 1953). The question in each case is whether or not there is a common mechanism on the cell surface which is responsible for the movement of both phosphate and sugar, or whether the correlation between the two is a consequence of events in the interior of the cell which follow the independent inward movement of these substances.

In the yeast cell, the evidence points to the presence, in the periphery of the cell, of a complete glycolytic system which is responsible for the uptake of sugar as well as for the uptake of phosphate. It should be noted, however, that if no extracellular phosphate is present, the sugar uptake proceeds at a somewhat reduced rate by recycling the phosphate available in the cell. If extracellular phosphate is present, it can enter the cycle, and be taken into the cell. Under no circumstances is there a stoichiometric relationship between phosphate uptake and glucose uptake. Theoretically, 2 mols of phosphate can be esterified for each mol of glucose used. Actually only $^1/_{10}$ of a mol of phosphate is taken up from the medium per mol of glucose under optimal conditions, indicating a maximal efficiency of only 5% in transport of phosphate (GOODMAN and ROTHSTEIN 1954). Thus it must be concluded that the primary event a the cell surface is the uptake of sugar, and that the operation of the sugar transport system can secondarily pick up extracellular phosphate.

In the yeast cell phosphate is taken up continuously as a requirement for the synthesis of new protoplasm during growth and cell division. In mammalian tissues such as muscle and red blood cells, there is no new growth. Therefore, phosphate is required primarily for maintenance of the phosphorylation cycles. Hence no dramatic alterations in the cellular content of phosphate occur as a consequence of metabolism of sugars, even though studies with $P^{32}$ labeled phosphate indicate a rapid transfer of the isotope into the cell in the presence of sugar (SACKS 1953). The $P^{32}$-transfer is primarily an exchange of cellular and extracellular phosphates, and not a phosphate uptake in the sense of an increased phosphate content of the cell. For this reason, the specific activity of the extracellular phosphate falls rather rapidly due to dilution with unlabeled cellular phosphate which is exchanging out.

Evidence has been presented which indicates that the equilibrium of extracellular $P^{32}$ with intracellular phosphate proceeds by way of phosphorylation reactions in the cell surface of red blood cells and of muscle. As in the yeast cell, these reactions are not operative in the absence of sugar, or do they have the primary function of increasing the phosphate content of the cell. It seems likely that the surface phosphorylation reactions are primarily concerned with the uptake of sugar; that reactions in sugar uptake at the periphery of the cell involve cycling of phosphate; and that during the operation of this cycle some of the $P^{32}$ labeled phosphate molecules of the medium exchange with peripheral phosphate of the cell and get carried into the cycle. This concept is supported by the fact that in red blood cells, the surface compounds that serve as precursors are 2, 3-diphosphoglycerate and ATP (GOURLEY 1952 a; PRANKERD and ALTMAN 1954), both associated with glycolytic reactions which can esterify phosphate. It is concluded that the studies of $P^{32}$-transfer support the concept that the transport of sugars is accomplished by enzymic phosphorylation reactions in the surface of the cell. In the case of muscle cells, because insulin increases the rate of $P^{32}$-transfer (SACKS 1952) as well as the rate of sugar uptake, it is entirely possible that the hormane stimulates the transport of sugar by acting on surface phosphorylation reactions.

## Transport of Amino Acids

Certain of the amino acids can be actively transported by cells and tissues. That is, they can be moved against the concentration gradient with the expenditure of energy by the cells. The uptake of amino acids has been studied in the intestine (HÖBER 1946), the kidney (SMITH 1951), red blood cells (CHRISTENSEN, RIGGS, and RAY 1952), microorganisms (GALE 1954), and tumor cells (CHRISTENSEN, RIGGS, RAFN, RAY, and PALATINE 1952). The most detailed studies have been carried out by GALE and co-workers reported in an extensive series of papers and reviewed recently (Gale 1954).

The mechanism by which active transport of amino acids is accomplished is not well understood. Gale (1954) discusses the active accumulation of glutamic acid by staphylococci in some detail. If amino acid deficient cells are suspended in glutamic acid, there is no accumulation of the amino acid. However, if a source of carbohydrate is added, such as glucose, a rapid accumulation takes place. The internal concentration of glutamic acid may rise to 400 times that of the external concentration. The accumulation has a high temperature coefficient and it is abolished by inhibitors which interfere with glucose metabolism. Thus there is no question that a metabolically linked transport is involved. Furthermore, the rate of uptake of glutamic acid is independent of its concentration over a wide range. Some component of the transport system is "saturated" by higher concentrations of glutamic acid. The transport system must involve a chemical combination of glutamic acid or one of its products with a limited quantity of carrier.

Gale discusses several possible explanations for the experimental observations:

(1) That enzymes in the cell membrane convert glutamic acid to a product which moves through the membrane and is reconverted to glutamic acid in the cell.

(2) That a membrane carrier system, coupled to glucose metabolism, combines with glutamic acid and carries it across the membrane.

(3) That glutamic acid diffuses through the membrane but is "trapped" in the cell by conversion to another form which analyzes as glutamic acid by existing analytical methods.

The first two hypotheses require enzymic activity at the cell surface; the last does not. The last hypothesis has been tested by using a number of different methods of breaking up the cell. In each case, specific methods for glutamic acid (an enzyme method and paper chromatography) give the same results. No evidence has been obtained that the internal glutamic acid is present in any form other than glutamic acid. On this basis, the third hypothesis seems untenable. If the first hypothesis is correct, then the penetrating form of glutamic acid is presumably some simple derivative, but no derivative yet tested meets the requirements of the hypothesis. It can be concluded that the exact mechanism of uptake of amino acids is as yet unknown, but it seems likely that surface enzyme activity is involved in some manner. If the surface enzymes do not alter the glutamic acid directly, then they probably supply the energy for a carrier-transport mechanism.

## Transport of Ions

It is becoming increasingly evident that the cell maintains its relatively constant internal ionic environment in the face of entirely different extra-cellular environment by means of a system of "ion pumps," located in the membrane of the cell and energized by metabolic reactions. It was not long ago that cell membranes were regarded to be relatively impermeable to ions. With the advent of radioactive isotopes it was found, however, that cells were in reality relatively permeable to certain ions. To explain how cells could exclude certain ions such as $Na^+$ and accumulate others such as $K^+$ in higher concentrations than present in the environment, it was suggested that cells are impermeable to $Na^+$ and that $K^+$ might be trapped in the cell by binding with a cellular constituent. Neither explanation is valid in the light of present information. Rather, the membrane is permeable to sodium, but this ion continuously expelled by a "sodium pump" in the membrane.

The study of ion movements is complicated because of the electrical charge involved. The movement of a given ion is not independent of the movement of other ions. If electrical neutrality is to be maintained a cation can only enter a cell (a) in exchange for a cation leaving the cell, (b) in company with an anion, or (c) to balance an internally produced anion. Thus it is sometimes difficult to determine which of the ions are being transported and which are moving in response to the electrical gradient. Many controversies have arisen because of this complication. In the case of the salt-respiration in plants, one group of workers favors an active transport of anions and a passive movement of cations, while a second group favors the active transport of cations with a passive movement of anions. An important consequence of ion transport is the electrical charge set up across the membrane, particularly in the polarization and conduction of nerve and muscle.

Ion transport has been intensively studied in many tissues. For details, the reader is referred to the following reviews: yeast (Conway 1954), brain and kidney (Davies 1954), plants (Lundegaardh 1954; Russell 1954; Steward and Miller 1954), red blood cells (Solomon 1952; Harris 1954; Maizels 1954), muscle (Conway 1954; Keynes and Hodgkin 1954; Steinbach 1954), nerve (Hodgkin 1951; Shanes 1952), an frog skin (Ussing 1949, 1954). It is generally agreed that ions are continually transported across the membranes of living cells, but the exact mechanism is a matter of speculation. A popular concept is that of a carrier system. The ion is assumed to form a specific complex with a substance in the lipid phase of the membrane. This complex then diffuses across the membrane and liberates the ion into the internal aqueous phase. Such a system will allow equilibration of the ion across the membrane but will not account for an active transport against the electrochemical potential. The transport must be accomplished by a coupling of the carrier system to metabolism. The exact nature of the carrier and of the coupling are at present in the realm of pure speculation. However, reasonable models have been described which involve surface enzymes capable of altering, synthesizing or destroying the carrier (Franck and Meier 1947; Rosenberg 1948; Osterhout 1952; Solomon 1952).

Another type of ion transport system which has been proposed involves the operation of redox systems, in which the creation of ionic charges associated with electron flow is coupled to ion movements (Conway 1951; Lundegaardh 1954). Such systems also require localization of enzymes in the cell surface, in this case redox enzymes such as the cytochromes.

It is also possible to devise a model which will pump ions but which does not involve surface enzymes. For example, a counterflow of a carrier substance produced inside the cell and flowing outward to the medium could pump out sodium. However, such a system would be inefficient, with a continuous loss of the carrier. Furthermore, the carrier substance could be detected in the medium. It is perhaps more reasonable to assume that the energy supplying reactions occur in the membrane itself.

## Surface Enzymes which Synthesize Extracellular Substances

A great many substances are secreted or liberated into the environment by living cells. The substances range from rather simple molecules to very complex macromolecules. In addition, cell walls, fibers and other solid structures may be elaborated by the living cell. In most cases, little is known about either the mechanisms or location of the synthesis of these materials. Three possibilities can be cited: (a) The materials can be synthesized inside the cell and then secreted in some manner; (b) Enzymes might be secreted into the environment which would synthesize the final product from reactants present in the extracellular space; (c) The products could be synthesized by enzymes located on the outer surface of the cell. Little attention has been given to the possible role of surface enzymes in the synthesis of extracellular substances and much of the following discussion is therefore based on inference.

### Proteins

Many extracellular proteins are found in animals. Examples include plasma proteins, digestive enzymes, and fibrous proteins such as silk, collagen and hair. It does not seem reasonable that proteins can be synthesized outside the cell by extracellular enzymes. Not only has extracellular protein synthesis never been observed, but the proteins are put together in such a specific and reproducible manner that specific templates, arays of enzymes, and a large number of animo acid substrates are undoubtedly required. If the proteins are synthesized inside the cell, how do they get out? It is not realistic to suppose that a protein molecule can simply diffuse through the cell membrane. Proteins are not lipid soluble and they are too large to pass through pores in the membrane. Any membrane with pores large enough to pass proteins would allow the leakage of all of the essential ions and low molecular weight compounds. Most cells do not leak proteins. In many of the cases in which extracellular proteins do appear, some specialized mechanism is involved. For example, in the mammalian organism, glandular cells are capable of excreting proteins, but a cytological examination of these cells reveals that a mechanical extrusion of a part of the cytoplasm or of granules or vacuoles can account for the protein secretion. The goblet cells of the intestinal mucosa form a large central vacuole which is extruded. The cells of the thyroid gland contain droplets of active colloid which migrate to the surface of the cell and are pinched off. In the pancreas, certain granules are present which form vacuoles which are expelled from the cell. Other examples of mechanical extrusion of secretory substances can be found in cytology texts. The extrusion of bulk material is a non-molecular form of secretion which cannot be evaluated by conventional physical chemical concepts.

Not all cases of protein secretion are associated with mechanical extrusion from the cytoplasm. Liver cells produce abundant quantities of plasma proteins with no evidence of mechanical extrusion. However, in this case, evidence obtained by MILLER and BLY (1951), MILLER, BLY, WATSON, and BALE (1950), and BLY, MILLER, and BALE (1950) supports the concept that the synthesis takes place on the surface of the cell. Using perfusion techniques and $C^{14}$ labeled amino acids, it was found that all of the plasma albumin and almost all of the alpha and beta globulins (including fibrinogen [MILLER and BALE 1954; MILLER, BLY, and BALE 1954]) were synthesized by the liver cells. A mixture of the essential amino acids, plus lysine-$\varepsilon$-$C^{14}$, was perfused through intact rat liver. $C^{14}$ appeared both in the synthesized plasma proteins and in the proteins and free amino acids of the liver cell. However, if a mixture of unlabeled, essential, amino acids was then perfused through the labeled liver, the resulting plasma protein was virtually unlabeled. Thus plasma proteins must be synthesized in a cellular location where no mixing could with the labeled amino acids already present in the liver cell. In other words, there is no obligate *intracellular* precursor of the circulating plasma proteins. These studies suggest the possibility that plasma protein synthesis occurs in the periphery of, or on the surface of the liver cell.

The hypothesis of a synthesis of plasma proteins on the surface of the cell is supported by other experiments (MILLER 1954). For example, if an unlabeled liver is perfused with labeled plasma protein, the distribution of the plasma protein is purely extracellular with no evidence that plasma proteins can pass through the cell membrane by any mechanism. In addition to the impermeability of liver cells to plasma proteins, there is good evidence that the liver cells are impermeable to its own cellular proteins. A liver that has been fed labeled amino acids, incorporates the activity into its cellular proteins. These proteins do not leak out of the cells. Similarly in an eviscerate carcass in which the cell proteins are labeled, it can be shown that there is no outward leakage of protein from the cells. The general impermeability of cells to protein, and in particular the impermeability of liver cells to the plasma proteins as well as its own cellular proteins, is incompatible with the concept of a plasma protein synthesis within the interior of the cell, but lends support to the concept of a synthesis on the cell surface.

Many microorganisms produce enzymes which appear in the medium. Lysis of cells may account for some of the protein. Synthesis of the protein may occur on the surface of the cell. Transport of the protein from the interior of the cell to the medium by some special mechanism cannot be excluded as a possibility, although no such mechanism has as yet been demonstrated.

The formation of extracellular protein fibers such as collagen, silk, hair, etc., is of some interest. There is no direct evidence concerning the site of formation of fibers. However, electron-micrographs of fiber producing systems are of some interest. For example, Watson (1954) has studied the odontoblasts of the developing tooth, in which collagen fibers are laid down in association with the polysaccharide, chondroitin sulfate. There is no evidence in this tissue, or in others, that the fibers are synthesized inside the cell and are then extruded. Rather, the appearance of the cells and fibers suggests that they are synthesized on the cell surface.

The formation of various jelly coats of marine eggs is of some interest. Runnström (1952) has discussed the problem in detail in relation to fertilization. The jelly coats are often composed of protein in combination with polysaccharides. The constituent sugars may be mixtures of glucose, fructose and galactose or of sulfonic acid derivatives. In some cases it is apparent that after fertilization, a membrane is formed partly from a preexisting vitelline membrane and partly from granules located in the cortex of the cell, directly under the plasma membrane.

## Carbohydrates

A large variety of extracellular carbohydrate polymers are synthesized by living cells. In some cases the polysaccharides are secreted into the environment, as for example the various slimes and mucins; in others the polysaccharides form closely adhering structures such as the cell walls of plants and the capsules of bacteria. The chemistry of the polysaccharides has been thoroughly investigated (Evans and Hibbert 1946; Berger 1950; Hehre 1951 b; Salton 1953; Whistler and Smart 1953). Some are relatively simple polymers of a single sugar such as: levans (fructose) and dextrans (glucose) produced by bacteria, mannans and glucans of the cell wall of yeast, and cellulose of plant cell walls. Others are complex polymers of two or more sugars. Some contain sugar acids, sugar amines, phosphate, sulfate and methyl esters. In the case of the bacterial capsules, the sugars may be arranged in a specific pattern possessing antigenic properties. The molecular weights may be as high as 2,000,000. Often, the polysaccharides are conjugated with proteins (mucopolysaccharides).

What is the site of synthesis of the various polysaccharides? If they are synthesized inside the cell, how do they get out, particularly in view of the impermeability of cells to sugars of low molecular weight? Of course it is possible that a mechanical extrusion of material from the cells, such as described for protein secretion may account for the mucins secreted by mammalian glands. But such a mechanism does not seem to be operative in other cells capable of producing extracellular polysaccharides.

Another possibility is the extracellular synthesis of the polysaccharides by excreted enzymes. Such might be the case with simple polysaccharides such as dextrans in which a single enzyme given a single substrate, sucrose, can produce the polymer, with no coupling to cell metabolism (HEHRE 1951 a). In other cases, however, metabolic energy of the cell is required. For example, certain yeasts form a starch-like capsule (MAGER and ASCHNER 1947). This capsule although it is a glucose polymer can be synthesized from other sugars. The cell must first convert the given sugar to glucose before incorporating it into the capsular material. In the case of the *Pneumococcus*, capsular material can be synthesized by a non-growing cell fed glucose, phosphate, Mg and K, if oxygen is present (BERNHEIMER 1953). The polymer that results is a complex one containing a number of different sugars in a specific array. It is inconceivable that the necessary inter-conversions and synthesis could be accomplished by secreted enzymes.

Some of the extracellular polysaccharides are undoubtedly synthesized by enzymes on the surface of the cell. For example, in a number of or-ganismus, B. subtilis, B. polymixa and Aerobacter levanicum, the levan pro-ducing enzyme, was not found in the medium, but only in the cells (HESTRIN, et al. 1943). In view of the fact that cells in general are impermeable to sucrose and in view of the high molecular weight of the levans formed, it is unlikely that they are produced inside the cell and diffuse out. It seems much more reasonable that the enzyme is bound on the surface of the cell and produces the levan at this site. In other organisms, the enzyme is found in the medium and can produce levan in the absence of the cells. It is perhaps secreted by the cells, or perhaps it is liberated by lysis of some of the cells in the growing population.

Cell wall materials are probably synthesized in situ by enzymes of the outer surface of the cell which are in direct contact with the cell wall. Cell walls of plant cells are laid down as cellulose fibers in a definite pattern of layers. There is no evidence that the fibers are formed inside the cell and mechanically transported to the surface. In fact, electron-micrographs suggest a local synthesis (FREY-WYSSLING 1952; PRESTON 1952; BROWN, REITH, and ROBINSON 1952). There is also some evidence that the plant cell surface possesses surface enzymes capable of hydrolyzing cellu-lose. For example, the cellulose content of certain tissues decreases in the absence of external substrate (BROWN and SUTCLIFFE 1950).

Capsules of bacteria are somewhat different from the rigid cell wall of plants. The capsular material constantly washes away and is resyn-thesized even in non-growing cell if a source of energy is supplied (BERN-HEIMER 1953). As already pointed out, the capsular polysaccharides are complex structures of high molecular weight. Unless the cell possesses some unknown mechanical means for extruding the polysaccharides from the interior of the cell, it must be concluded that the polymers are synthesized on the surface of the cell. STACEY (1949) has discussed the relationship of polysaccharide synthesis to nucleic acids of the cell surface in some detail. He believes that desoxyribonucleo-proteins of the cell surface together with the necessary "prosthetic groups and coenzymes, constitute the various enzyme systems responsible for cell functions and particularly for the build-up and breakdown of the cell macromolecules."

## Surface Enzymes which Maintain and Synthesize the Cell Surface Structure Itself

The surface of the cell is a structure with specific properties which vary from cell to cell. The complexity of the surface structure is evident from its mechanical, permeability, chemical and electrical properties. The cell surface possesses enzymes and systems for transporting or pumping various substances into and out of the cell. It possesses specific antigenic properties which differ from cell to cell. Chemically speaking, the cell surface has been shown to contain a variety of proteins, lipids, phospho-lipids, polysaccharides, nucleic acids. For example, the red cell stroma contains a number of specific proteins and lipid fractions (MOSKOWITZ and CALVIN 1952). Nucleic acid in the surface is responsible for gram-positive staining in bacteria (STACEY 1949) and for the binding of bivalent ions (LANSING and ROSENTHAL 1952). Streptomycin apparently acts on the surface of the cell by binding to nucleic acid, as indicated by the effect on electrophoretic mobility of the cell (McQUILLEN 1951). Immunological studies indicate that the surface of bacteria contains polysaccharide as well as protein and phospholipids (MILES and PIRIE 1949). MITCHELL and MOYLE (1951) found that the cell envelope of bacteria contains a glycerophospho-protein complex. Many other examples of the chemical and structural complexity of the surface could be mentioned.

In a growing and dividing cell, the cell surface material must increase. Synthesis of each cell surface material must take place. Where does the synthesis occur? Are the various components synthesized inside the cell and then transported to the periphery where they somehow combine or arrange themselves in the correct manner and sequence by non-enzymatic reactions? Or are they synthesized directly at the site they will occupy by enzymes present in the cell surface? At the present time we have little evidence bearing directly on this problem. The concepts presented here are therefore speculative. There are a number of observations which are compatible with the hypothesis of a synthesis, at least in part, of materials at the cell membrane itself, rather than in the interior of the cell. In the first place the cell membrane is a definite structure. The structural components are fixed in place and are not free to diffuse away. The proteins, phospholipids and other components of the structure, if they were synthesized inside the cell, would have to find their way to their correct locations in the membrane and would have to join together in a specific orientation in order to account for the specific properties of the membrane such as its permeability and antigenic properties. It seems improbable that the proper arrangements of the building blocks can occur by any sort of random movements starting at the inside of the cell. Not only would each of the many kinds of building blocks have to arrive at the correct place, but they would have to combine chemically in the various configurations which as a whole would comprise the membrane. It seems more likely that the building blocks are "cemented in place" by reactions occurring in the membrane. Perhaps some of the building blocks themselves, such as structural proteins, are also synthesized in the membrane.

What evidence can be mobilized to support the concept of a synthesis of cell surface structure at its final site? Recently, ALTMAN, WATMAN and SOLOMON (1951) and ALTMAN (1953) have studied the incorporation of labeled acetate into the lipid fractions of the red cell stroma. Turnover was very rapid (half life of 3 to 4 days) in the sphingolipid fraction, in cholesterol and in the phosphatids. With the latter compounds, experiments were performed with $P^{32}$ labeled phosphate as well as with labeled acetate. The incorporation of $C^{14}$ and $P^{32}$ into these compounds can not arise from exchange but only from synthetic processes. It must therefore be concluded that the lipid components of the red cell membrane are constantly degraded and resynthesized by enzymes located in the stroma. The structural elements of the membrane are in a dynamic, not a static, state.

There is evidence that the formation of adaptive enzymes may involve the synthesis of proteins in the surface of the cell. For example, many yeasts cannot immediately utilize galactose (LINDEGREN and PALLERONI 1952). Nevertheless, a few hours after such cells are exposed to galactose in the presence of glucose (but no nitrogen source), the cells develop the ability to ferment galactose, that is, they "adapt" to galactose (SPIEGELMAN, REINER, and COHNBERG 1947). The adaptive process has been studied in some detail from both a genetic and a biochemical point of view. It is associated with the appearance of an enzyme activity which was not present in measureable amounts in unadapted cells. Thus, zymase preparations from unadapted yeast cannot ferment galactose, whereas a similar preparation from adapted cells will ferment this sugar (SPIEGELMAN, REINER, and MORGAN 1947). The enzyme in this case is galactokinase, which converts galactose to galactose-1-phosphate in the presence of ATP (WILKINSON 1949). The appearance of an enzyme activity does not necessarily mean that a new enzyme has been synthesized. The enzyme may be present in unadapted cells, but in an inactive form. Although the exact mechanism of enzymic adaptation is not known, there is evidence that a synthesis of new protein may be involved. HALVORSON and SPIEGELMAN (1953) found that during adaptation to maltose in yeast, there was a marked diminution in the free amino acid content of the cell. Furthermore, the presence of a source of nitrogen, markedly stimulates the adaptive process (PORTER, HOLMES, and CROCKER 1953).

From the point of view of the cell surface, it is of considerable interest that many of the substrates which involve an adaptive response, do not ordinarily penetrate into the cell at an appreciable rate. Certain of the non-penetrating substrates have already been discussed such as galactose, sucrose, maltose and lactose. Another substrate which presumably does not penetrate is colloidal starch, which induces amylase formation in *Streptococcus bovis* (RAHN and LEET 1949).

Some of the adaptive enzymes such as the disaccharide-splitting enzymes (invertase, lactase and maltase) are located on the surface of yeast cells (see previous section). Each of these enzymes is adaptive in certain strains or species of yeast. Although the adaptive strains have not been specifically tested for surface enzymes, it is probable that they resemble the strains that have been tested in this respect.

Surface reactions have been investigated to some extent in galactose-adapted cells. During adaptation the membrane is altered from one through which galactose cannot penetrate at an appreciable rate (CONWAY and DOWNEY 1950 a) into one through which galactose can be rapidly taken up. Concurrently, galactokinase activity appears (SPIEGELMAN, REINER, and MORGAN 1947). Studies of the inhibition of galactose uptake in adapted cells by uranium (ROTHSTEIN, MEIER, and HURWITZ 1951) suggest that the two events are closely linked,—that the ability of the cell to take up galactose is associated with the appearance of galactokinase in the cell surface.

Admittedly, the evidence presented above is incomplete and not unequivocal. Nevertheless, it supports the concept that during adaption to certain non-penetrating substrates, synthesis, partial synthesis, or at least activation of enzymes occurs in the surface of the cell.

Another phenomenon which is suggestive of a synthesis, or at least of an alteration of cell surface structure, is the phenomenon of "transformation" (see recent review of AUSTRIAN [1952]). Pneumococci of one type produce a specific polysaccharide capsule. If treated with a desoxyribonucleic acid preparation from a second type of organism, a few cells in the population are "transformed" into organisms which will produce the type of polysaccharide associated with the strain donating the factor. There is considerable evidence that the transforming factor is desoxyribonucleic acid (ZAMENHOF, ALEXANDER, and LEIDY 1953). It has already been pointed out that the polysaccharide capsule is probably synthesized on the outer surface of the cell, by an array of enzymes. STACEY (1949) suggests that desoxyribonucleic acids in the cell surface are an integral part of the enzymic systems for synthesizing the polysaccharides, and that the transforming factor simply alters the enzymic array in the membrane, resulting in the formation of a new type of polysaccharide.

One important aspect of adaptive enzyme formation and of transformation has not yet been discussed. It concerns the mechanism by which the genes, located in the nucleus, control the synthesis of structures or enzymes at the surface of the cell. The ability of cells to adapt is a gene-controlled property (LINDEGREN, SPIEGELMAN, and LINDEGREN 1944). If it is assumed that the adaptive enzyme is synthesized at the cell surface in response to non-penetrating substrates, then the cell surface itself must contain the template which in conjunction with the proper substrate produces the enzyme, with no direct intervention of the gene itself. There is some evidence that this is the case. In crosses of adapting and non-adapting yeast cells, followed by segregation, the adaptive enzymes can be maintained even in those lines which do not possess the requisite genes, if the substrate to which the yeast is now adapted is continuously present (SPIEGELMAN, LINDEGREN, and LINDEGREN 1945). The adaptive enzyme disappears permanently in these lines if the special substrate is absent. Therefore, it is possible that the cell membrane contains a template for adaptive enzyme formation, which is replicable under appropriate conditions, independent of genic control. In the case of "transformation," a permanent change is induced in the organisms, which is maintained in the following generations, even though exogenous transforming factor is no longer present. This phenomenon can also be interpreted as indicating that a replicable, transformable structure in the cell surface serves as a template for polysaccharide synthesis.

There has been considerable interest on the part of geneticists in the problem of replicable cytoplasmic structures such as plastids and killer factor (kappa) in *Paramecium*. Evidence also points to replication in the case of mitochondria and other granules. SONNEBORN (1951) has recently obtained evidence that certain constituents of the cell surface are also replicable, with relatively little intervention by the genic complement in some cases. He studied antigenic properties of the cell surface of *Paramecium* as they are influenced by environmental conditions (temperature) and by genic factors. He found that different genes of the same nucleus "may be capable of determining two or more alternative, mutually exclusive traits, only one of which can come to full phenotypic expression in any one cell. The decision is binding during subsequent cell multiplication for a shorter or longer time, and under certain conditions, permanently. Thus two cells with the same genes and under the same conditions may have different alternative traits and maintain them during cell reproduction." SONNEBORN's studies indicate that certain elements of the cell surface can perhaps be replicated independently of specific genic control.

The existence of replicable structures in the cell surface is compatable with the concept that synthesis of new cell surface structure during growth is mediated by arrays of enzymes located in the structure itself. The general pattern of the enzymic array perhaps constitutes the template which determines the nature of the end product. The template is under genic control, but it is also influenced by external factors such as temperature and the availability of substrates. The nature of the genic control over the patterns of enzymic arrays in a replicable structure is a complex problem beyond the scope of the present paper.

## General Conclusions

Existing information concerning the enzymology of the cell surface is fragmentary at best. However, sufficient information is available to indicate the overall general pattern. The surface of the cell is not a static structure or a "skin." It participates actively in many cellular functions, playing an important part in many interactions between the cell and its environment. Four biochemical functions can be attributed to the cell surface: (1) digestion of non-utilizable substances of the environment into useable products, (2) synthesis of extracellular structures and diffusible macromolecules, (3) active transport and secretion of substances either by altering these substances in the membrane, or by providing energy for their transport by "carrier mechanisms", and (4) self-maintenance and perhaps replication of new surface during growth. In some cases unequivocal proof of the existence of surface enzymes has been presented; in others, reasonable interpretations have been made. Much work remains to be done on all aspects of the problem to verify or correct the hypotheses, to add more details and perhaps to demonstrate new functions of the cell surface. It is hoped that the present monograph will provide stimulus for further work, or at least, that it will bring to those who work with cells an awareness of the enzymic activities of the cell surface.

The fact that the cell surface participates in metabolic reactions has important implications for the toxicologist, the pharmacologist, the endocrinologist, as well as the physiologist and biochemist. Any substance placed in the environment of a cell must equilibrate with the cell surface before it penetrates into the interior. The first activities of the cell which can be influenced by the substance are the cell surface reactions. Substances which cannot penetrate into the interior of the cell can *only* act on the cell surface. Substances which *can* penetrate into the cell interior may influence either surface or internal reactions, or both. For example, uranyl ion has a marked affinity for certain anionic groups. When cells are exposed to uranium, the metal is bound by the anionic groups of the cell surface and it can penetrate into the interior of the cell only at a low rate. The predominant, acute, toxic action of uranium is therefore on cell surface enzyme reactions. In the case of mercury, some of the metal is bound by the cell surface, resulting in an inhibition of surface enzyme systems, but some also penetrates into the interior of the cell resulting in inhibition of other cellular functions. The surface effect is rapid and occurs with lower concentrations (Demis 1953).

In the case of endocrine substances or humoral agents, it has been shown that insulin may act by binding to the surface of the cell. If insulin acts on the cell surface, then adrenalin may also act on the cell surface, because adrenalin counteracts the effect of insulin on isolated muscle. The parasympathetic substance, acetyl choline, seems to act at the surface of the striated muscle cell and at the membranes of the ganglionic synaptic junctions. Also the enzyme that destroys acetyl choline is located on the surface of cells. Many of the autonomic drugs may act on surface reactions by competition with or substitution for acetyl choline and adrenalin. Clark (1937) first pointed to be possibility that drugs might act on the cell surface. Unfortunately, too little attention has been given to this aspect of Pharmacology.

The presence of enzyme in the cell surface structure, as well as in the mitochondria, the nucleus, the granules, and the other sub-cellular elements, emphasize the importance of cellular architecture in cellular metabolism and function. The biochemical and physiological activities of cells are the result of the organized activities of groups of enzymes associated with cellular structures, rather than the summation of the activities of individual enzymes.

## Acknowledgment

The author wishes to express his gratitude to his many colleagues and friends who patiently tolerated his questions and his polemics, and who contributed their time, their advice and their encouragement. A special bouquet is due those hardy souls who undertook the arduous task of reading the manuscript to eliminate errors of omission, commission and confusion: K. Altman, H. L. Berke, K. W. Cooper, I. Feldmann, H. C. Hodge, L. Hurwitz, M. Marlett, L. L. Miller, W. Neuman, M. Rothstein, N. S. Stannard.

This report is based in part on work performed under contract with the U. S. Atomic Energy Commission at the University of Rochester Atomic Energy Project.

## Bibliography

Alivisatos, S. G. A., and O. F. Denstedt, 1951: Lactic Dehydrogenase and DPN-ase Activity of Blood. Science 14, 281–283.

Altman, K. I., 1953: The *in vitro* Incorporation of a-C¹⁴-Acetate into the Stroma of the Erythrocyte. Arch. Biochem. Biophys. 42, 478–480.

— 1954: Unpublished observation.

— R. N. Watman, and K. Saloman, 1951: The Incorporation of a-C¹⁴-Acetate into the Stroma of the Erythrocyte. Arch. Biochem. Biophys. 33, 168–169.

Austrian, R., 1952: Bacterial Transformation Reactions. Bact. Rev. 16, 31–50.

Barany, E., and E. Sperber. 1939: Absorption of Glucose Against a Concentration Gradient by the Small Intestine of the Rabbit. Skand. Arch. Physiol. 81, 290–299.

Barron, E. S. G., M. I. Ardao, and M. Hearon, 1950: Regulatory Mechanism of Cellular Respiration. III. Enzyme Distribution in the Cell. Its Influence on the Metabolism of Pyruvic Acid by Baker's Yeast. J. gen. Physiol. 34, 211–224.

— J. A. Muntz, and B. Gasvoda, 1948: Regulatory Mechanisms of Cellular Respiration. I. The Role of Cell Membranes: Uranium Inhibition of Cellular Respiration. J. gen. Physiol. 33, 163–178.

Berger, L., M. W. Slein, S. P. Colowick, and C. F. Cori. 1946: Isolation of Hexokinase from Baker's Yeast. J. gen. Physiol. 29, 379–391.

Berke, H., and A. Rothstein, 1954: Unpublished observations.

Bernheimer, A. W., 1953: Synthesis of Type III Pneumococcal Polysaccharide by Suspensions of Resting Cells. J. exper. Med. 97, 591–600.

Bly, C. G. L. L. Miller, and W. F. Bale, 1950: Minor Role of Non-hepatic Tissues in Plasma Protein Synthesis Observed with the Aid of Lysine-ε-C¹⁴. Fed. Proc. 9, 153.

Boell, E. J., and D. Nachmansohn, 1940: Localization of Choline Esterase in Nerve Fibers. Science 92, 513.

Booy, H. L., 1940: The Protoplasmic Membrane Regarded as a Complex System. Rec. Trav. Botan. Neerland. 37, 1–77.

Bourne, G., 1943–44: The Distribution of Alkaline Phosphatase in Various Tissues. Quart. J. exper. Physiol. 32, 1–19.

Bradfield, J. R. G., 1949: Phosphatase Cytochemistry in Relation to Protein Secretion. Exper. Cell. Res. Suppl. 1, 338–350.

— 1950: The Localization of Enzymes in Cells. Biol. Revs. 25, 113–157.

Brauer, R. W., and M. A. Root, 1945: The Cholinesterase of Human Erythrocytes. Fed. Proc. **4**, 113.

Brooks, S. C., and M. M. Brooks, 1941: The Permeability of Living Cells. Protoplasma-Monographien 19.  Berlin-Zehlendorf, Verlag Gebr. Borntraeger.

Brown, R., 1952: Protoplast Surface Enzymes and Absorption of Sugar. Internat. Rev. Cytology **1**, 107–118.

— and J. F. Sutcliffe, 1950: Effects of Sugar and Potassium on Extension Growth in the Root. J. exper. Bot. **1**, 88–113.

— W. S. Reith, and E. Robinson, 1952: The Mechanism of Plant Cell Growth. Symp. Soc. exper. Biol. **6**, 329–347.

Burger, M., 1950: Bacterial Polysaccharides: Their Chemical and Immunological Aspects.  Charles C. Thomas, Springfield, Ill.

Burstrom, H., 1941: Studies on the Carbohydrate Nutrition of Roots. Ann. Agr. Coll. Sweden **9**, 264.

Campbell, P. N., and H. Davson, 1948: Absorption of 3-Methylglucose from the Small Intestine of the Rat and the Cat. Biochem. J. **43**, 426–429.

Causey, G., and E. J. Harris, 1951: The Uptake and Loss of Phosphate by Frog Muscle. Biochem. J. **49**, 176–183.

Christensen, H. N., T. R. Riggs, M. L. Rafn, N. E. Ray, and I. M. Palatine, 1952: Concentrative Uptake of Amino Acids by the Ehrlich Mouse Ascites Carcinoma Cells.  J. biol. Chem. **194**, 57–68.

— — and N. E. Ray, 1952: Concentrative Uptake of Amino Acids by Erythrocytes *in Vitro.*  J. biol. Chem. **194**, 41–51.

Clark, A. J., 1937: General Pharmacology.  J. Springer, Berlin.

Clarkson, E. M., and M. Maizels, 1952: Distribution of Phosphatases in Human Erythrocytes.  J. Physiol. **116**, 112–128.

Conway, E. J., 1951: The Biological Performance of Osmotic Work.  A Redox Pump.  Science **113**, 270–273.

— 1954: Some Aspects of Ion Transport Through Membranes.  Symp. Soc. exper. Biol. **8** (in press).

— and M. Downey, 1950 a: An Outer Metabolic Region of the Yeast Cell.  Biochem. J. **47**, 347–355.

— — 1950 b: pH Values of the Yeast Cell.  Biochem. J. **47**, 355–360.

Cook, R. P., 1926: The Antagonism of Acetyl Choline by Methylene Blue.  J. Physiol. **62**, 160–165.

Cori, C. F., 1925: The Rate of Absorption of Hexoses and Pentoses from the Intestinal Tract.  J. biol. Chem. **66**, 691—715.

— 1931: Mammalian Carbohydrate Metabolism.  Physiol. Rev. **11**, 143—275.

— 1945–46: Enzymatic Reactions in Carbohydrate Metabolism.  Harvey Lectures **41**, 253–272.

— and G. T. Cori, 1929: Fate of Glucose and other Sugars in the Eviscerated Animal.  Proc. Soc. exper. Biol. a. Med. **26**, 432.

Couteaux, R., and D. Nachmansohn, 1938: Cholinesterase at the end-plates of Voluntary Muscle after Nerve Degeneration.  Nature **142**, 481.

Cramer, F. B., and G. E. Woodward, 1952: 2-Desoxy-d-Glucose as an Antagonist of Glucose in Yeast Fermentation.  Notes from Biochemical Research Foundation **253**, 354–360.

Crane R. K., and A. Sols, 1953: The Association of Hexokinase with Particulate Fractions of Brain and other Tissue Homogenates.  J. biol. Chem. **203**, 273–292.

Csaky, T. Z., 1953: Phosphorylation of 3-Methylglucose by Hexokinase from Rat's Intestinal Mucosa.  Science **118**, 253–254.

Danielli, J. F., 1943: The Theory of Penetration of a Thin Membrane. Appendix A, 341–352, in Davson, H., and J. F. Danielli, The Permeability of Natural Membranes.  The University Press, Cambridge.

— 1951: Physical and Physiochemical Studies of Cells.  Part I. General, page 28. The Cell Surface and Cell Physiology.  Chapter III, page 68.  In Cytology and Cell Physiology, edited by G. Bourne, Clarendon Press, Oxford.

— 1952: Structural Factors in Cell Permeability.  Symp. Soc. exper. Biol. **6**, 1–15.

— 1953: Cytochemistry: A Critical Approach.  Wiley, New York.

Davies, R. E., 1954: Relations Between Active Transport and Metabolism in Some Isolated Tissues and Mitochondria.  Symp. Soc. exper. Biol. **7** (in press).

DAVSON. H., 1952: A Textbook of General Physiology. Chapter XII, page 297. The Blakiston Company, Philadelphia.

— and J. F. DANIELLI, 1943: The Permeability of Natural Membranes. University Press, Cambridge.

— and W. S. DUKE-ELDER, 1948: The Distribution of Reducing Substances Between the Intra-ocular Fluids and Blood Plasma, and the Kinetics of Penetration of Various Sugars into These Fluids. J. Physiol. 107, 141–152.

DEMIS D. J., 1953: A Study of the Effects of Insulin and of Mercury on the Utilization of Monosaccharides by Excised Rat Diaphragm. Ph. D. Thesis, University of Rochester.

— and ROTHSTEIN, 1954: The Absence of an Insulin Effect on Anaerobic Uptake of Glucose. Amer. J. Physiol. (in press).

— — and R. MEIER, 1954: The Relationship of the Cell Surface in Metabolism. X. The Location and Function of Invertase in the Yeast Cell. Arch. Biochem. Biophys. 48, 55–62.

DEMPSEY, E. W., and H. W. DEANE, 1946: The Cytological Localization, Substrate Specificity and pH Optima of Phosphatases in the Duodenum of the Mouse. J. cellul. a. comp. Physiol. 27, 159—171.

DENZ, F. A., 1953: On the Histochemistry of the Myoneural Junctions. Brit. J. exper. Path. 34, 329–339.

DERRICK, M. J., R. E. MILLER and M. G. SEVAG, 1953: Yeast Phosphatase. Immunological Method of Determining its Location in the Cell. Fed. Proc. 12, 196.

DORMER, K. J., and H. E. STREET, 1949: The Carbohydrate Nutrition of Tomato Roots. Ann. Bot. 13, 199–217.

DOUDOROFF, M., 1951: The Problem of the "Direct Utilization" of Disaccharides by Certain Microorganisms. In a Symposium on Phosphorus Metabolism, vol. 1, page 42, Part I. Johns Hopkins Press, Baltimore.

DRATZ, A. F., and P. HANDLER, 1952: Renal Phosphate and Carbohydrate Metabolism Studied with the Aid of Radiophosphorus. J. biol. Chem. 197, 419–431.

DRURY, D. R., and A. N. WICK, 1951: Insulin and the Volume of Distribution of Glucose. Amer. J. Physiol. 166, 159–164.

EGGLESTON, M. G., 1933: Diffusion of Inorganic Phosphate into and out of the Skeletal Muscles and Bones of the Frog. J. Physiol. 79, 31–48.

EISENMAN, A. J., L. OTT, P. K. SMITH, and A. W. WINKLER. 1940: A Study of the Permeability of Human Erythrocytes to Potassium, Sodium, and Inorganic Phosphate by the use of Radioactive Isotopes. J. biol. Chem. 135, 165–173.

EMMEL, V. M., 1946: The Intracellular Distribution of Alkaline Phosphatase Activity Following Various Methods of Histologic Fixation. Anat. Rec. 95, 159–175.

ENNOR, A. H., and H. ROSENBERG, 1954 a: An Investigation of the Turnover Rates of Organophosphates. I. Extracellular space and Intracellular Inorganic Phosphate in Skeletal Muscle. Biochem. J. 56, 302—307.

— — 1954 b: An Investigation of the Turnover Rates of Organophosphates. II. The Rate of Incorporation of $P^{32}$ into Adenosine Triphosphate and Phosphocreatine in Skeletal Muscle. Biochem. J. 56, 308–316.

EVANS, T. H., and H. HIBBERT, 1946: Bacterial Polysaccharides. Adv. Carbohydrate Chem. 2, 203–233.

FRANCK, J., and J. E. MAYER, 1947: An Osmotic Diffusion Pump. Arch. Biochem. 14, 297–313.

FREY-WYSSLING, A., 1952: Growth of Plant Cells. Symp. Soc. exper. Biol. 6, 320–328.

FURCHGOTT, R. F., and E. SHORR, 1943: Phosphate Exchange in Resting Cardiac Muscle as Indicated by Radioactivity Studies. IV. J. biol. Chem. 151, 65–86.

GALE, E. F., 1954: The Accumulation of Amino-Acids within Staphylococcal Cells. Symp. Soc. exper. Biol. 8 (in press).

GEMMILL, C. L., and L. HAMMAN, 1941: The Effect of Insulin on Glycogen Deposition and on Glucose Utilization by Isolated Muscles. Bull. Johns Hopkins Hospl. 68, 50–57.

GOLDSTEIN, M. S., and W. L. HENRY, B. HUDDLESTUN, and R. LEVINE, 1953: Action of Insulin on Transfer of Sugars Across Cell Barriers: Common Chemical Configuration of Substances Responsive to Action of the Hormone. Amer. J. Physiol. 173, 207–211.

Gomori, G., 1939: Microtechnical Demonstration of Phosphatase in Tissue Sections. Proc. Soc. exper. Biol. a. Med. 42, 23–26.

Goodman, J. W., and A. Rothstein, 1954: The Mechanism of Uptake of Phosphate bz Yeast Cells. Fed. Proc. 13, 57.

Gourley, D. R. H., 1952 a: The Role of Adenosine Triphosphate in the Transport of Phosphate in the Human Erythrocyte. Arch. Biochem. Biophys. 40, 1–12.

— 1952 b: Glycolysis and Phosphate Turnover in the Human Erythrocyte. Arch. Biochem. Biophys. 40, 13–19.

— and C. L. Gemmill, 1950: The Effect of Temperature upon the Uptake of Radioactive Phosphate by Human Erythrocytes in Vitro. J. cellul. a. comp. Physiol. 35, 341–352.

— and J. T. Matschiner, 1953: Rates of Exchange of Phosphate Ions in Human, Rabbit, and Chicken Blood. J. cellul. a. comp. Physiol. 41, 225–236.

Greenberg, D. M., 1952: The Acid Soluble Phosphates in Animal Metabolism. Vol. II, Chapt. I, pages 3–28, in A Symposium on Phosphorus Metabolism, edited by W. D. McElroy and B. Glass, Johns Hopkins Press, Baltimore.

Greig, M. E., J. S. Faulkner, and T. C. Mayberry, 1953: Studies on Permeability. IX. Replacement of Potassium in Erythrocytes During Cholinesterase Activity. Arch. Biochem. Biophys. 43, 39–47.

— and W. C. Holland, 1949: Studies on the Permeability of Erythrocytes. I. The Relationship between Cholinesterase Activity and Permeability of Dog Erythrocytes. Arch. Biochem. 23, 370–384.

— — 1951: Studies on the Permeability of Erythrocytes. IV. Effect of Certain Choline and Non-Choline Esters on Permeability of Dog Erythrocytes. Amer. J. Physiol. 164, 423–427.

Haft, D. E., and I. A. Mirsky, 1952: Action of Alloxan, Iodoacetate, and p-Chloro-Mercuribenzoate on Carbohydrate Metabolism of the Isolated Rat Diaphragm in the Absence and Presence of Insulin. J. Pharmacol. exper. Therap. 104, 340–347.

— — and G. Perisutti, 1953: Influence of Insulin on Uptake of Monosaccharides by the Isolated Rat Diaphragm. Proc. Soc. exper. Biol. a. Med. 82, 60–62.

Halvorson, H. O., and S. Spiegelman, 1953: Net Utilization of Free Amino Acids During the Induced Synthesis of Maltozymase in Yeast. J. Bact. 66, 601–608.

Harris, E. J., 1953: Phosphate Liberation from Isolated Frog Muscle. J. Physiol. 122, 366–370.

— 1954: Linkage of Na and K transport in Human Erythrocytes. Symp. Soc. exper. Biol. 8 (in press).

— and L. B. Gehrsitz, 1949: The Movement of Monosaccharides into and out of the Aqueous Humor. Amer. J. Ophthal. 32, 167–175.

Hehre, E. J., 1951 a: The Synthesis of Polysaccharides without the Intermediation of Phosphate. Vol. 1, pages 48–52, in Phosphorus Metabolism, edited by W. D. McElroy and B. Glass. Johns Hopkins Press, Baltimore.

— 1951 b: Enzymic Synthesis of Polysaccharides. Adv. Enzymol. 11, 297–337.

Heilbrunn, L. V., 1952: An Outline of General Physiology. W. B. Saunders Co., Philadelphia.

Hele, M. P., 1950: Phosphorylation and Absorption of Sugars in the Rat. Nature 166, 786–787.

Herbert, E., 1952: A Study of the Liberation of Orthophosphate from Adenosine Triphosphate by the Stromata of Human Erythrocytes. Ph. D. Thesis, Univ. of Pennsylvania.

Hestrin, S., S. Avineri-Shapiro, and M. Aschner, 1943: The Enzymic Production of Levan. Biochem. J. 37, 450–456.

Hevesy, G., 1947: Some Applications of Radioactive Indicators in Turnover Studies. Adv. in Enzymol. 7, 111–214.

— K. Linderström-Lang, and N. Nielsen, 1937: Phosphorus Exchange in Yeast. Nature 140, 725.

— and K. Zehran, 1946: The Effect of Röntgen Rays and Ultraviolet Radiation on the Permeability of Yeast. Acta Radiol. 27, 316–327.

Höber, R., 1946: Physical Chemistry of Cells and Tissues. The Blakiston Co., Philadelphia.

Hodgkin, A. L., 1951: The Ionic Basis of Electrical in Nerve and Muscle. Biol. Rev. 26, 339–409.

HOLLAND, W. C., and M. E. GREIG, 1950: Studies on Permeability. II. The Effect of Acetyl Choline and Physostigmine on the Permeability to Potassium of Dog Erythrocytes. Arch. Biochem. **26**, 151—155.

HOPKINS, R. H., and R. H. ROBERTS, 1935: The Kinetics of alcoholic Fermentation of Sugars by Brewer's Yeast. I. The Effect of Concentrations of Yeast and Sugar. Biochem. J. **29**, 919–936.

HOTCHKISS, R. D., 1943: Gramicidin, Tyrocidine and Tyrothricin. Adv. in Enzymol. **4**, 153–199.

HURWITZ, L., 1953: A Comparative Study of the Inhibitory Actions of Mercury and Uranium on Yeast and Yeast Hexokinase. Ph. D. Thesis, University of Rochester.

— and A. ROTHSTEIN, 1951: The Relationship of the Cell Surface to Metabolism. VII. The Kinetics and Temperature Characteristics of Uranium-Inhibition. J. cellul. a comp. Physiol. **38**, 437–450.

JEENER, R., and J. BRACHET, 1944: Recherches sur L'Acide Ribonucleique des Levures. Enzymol. **11**, 222–234.

JUNI, E., M. D. KAMEN, J. M. REINER, and S. SPIEGELMAN, 1948: Turnover and Distribution of Phosphate Compounds in Yeast Metabolism. Arch. Biochem. **18**, 387–408.

KALCKAR, M. H., J. DEHLINGER, and A. MEHLER, 1944: Rejuvenation of Phosphate in Adenine Nucleolides. II. The Rate of Rejuvenation of Labile Phosphate Compounds in Muscle and Liver. J. biol. Chem. **154**, 275–291.

KAMEN, M. D., and S. SPIEGELMAN, 1948: Studies on the Phosphate Metabolism of Some Unicellular Organisms. Cold Spring Harbor Symp. on Quant. Biol. **13**, 151–163.

KEYNES, R. D., and A. L. HODGKIN, 1954: Movements of Cations During Recovery in Nerve. Symp. Soc. exper. Biol. **8** (in press).

KJERULF-JENSEN, K., and E. LUNDSGAARD, 1940: Quantitative Wertung des Umsatzes der Phosphatester in der Darmschleimhaut von Ratten während der Fructoseresorption. Z. physiol. Chem. **266**, 217–224.

KOELLE, G. B., and J. S. FRIEDENWALD, 1949: A Histochemical Method for Localizing Cholinesterase Activity. Proc. Soc. exper. Biol. a. Med. **70**, 617–622.

KRAHL, M. E., 1951: The Effect of Insulin and Pituitary Hormones on Glucose Uptake in Muscle. Ann. N. Y. Acad. Sci. **54**, 649–670.

KUNITZ, M., and M. R. McDONALD, 1946: Crystalline Hexokinase (heterophosphatese) Method of Isolation and Properties. J. gen. Physiol. **29**, 393–412.

LANSING, A. I., and T. B. ROSENTHAL, 1952: The Relation Between Ribonucleic Acid and Ionic Transport Across the Cell Surface. J. cellul. a. comp. Physiol. **40**, 337–346.

LeFEVRE, P. G., 1948: Evidence of Active Transfer of Certain Non-Electrolytes Across the Human Red Cell Membrane. J. gen. Physiol. **31**, 505–527.

— 1954: The Evidence for Active Transport of Monosaccharides Across The Red Cell Membrane. Symp. Soc. exper. Biol. **8** (in press).

LEVINE, R., M. S. GOLDSTEIN, B. HUDDLESTUN, and S. P. KLEIN, 1950: Action of Insulin on the "Permeability" of Cells to Free Hexoses, as Studied by its Effect on the Distribution of Galactose. Amer. J. Physiol. **163**, 70–76.

LIBET, B., 1948: Adenosinetriphosphate (ATP-ase) in Nerve. Fed. Proc. **7**, 72.

LINDBERG, O., 1950: On Surface Reactions in the Sea Urchin Egg. Exper. Cell. Research. **1**, 105–114.

LINDEGREN, C. C., and N. J. PALLERONI, 1952: Absence of the Pre-adaptive Utilization of Galactose by Yeasts. Nature **169**, 879.

— S. SPIEGELMAN, and G. LINDEGREN, 1944: Mendelian Inheritance of Adaptive Enzymes in Yeast. Proc. Nat. Acad. Sci. **30**, 346–352.

LINDVIG, P. E., M. E. GREIG, and S. W. PETERSON, 1951: Studies on Permeability. V. The Effects of Acetylcholine and Physostigmine on the Permeability of Human Erythrocytes to Sodium and Potassium. Arch. Biochem. **30**, 241–250.

LUKENS, F. D. W., 1953: The Influence of Insulin on Protein Metabolism. Abstract XIXth Internat. Physiol. Cong., page 12.

LUNDEGARDH, H., 1954: Active Absorption and Transport of Ions. Symp. Soc. exper. Biol. **7** (in press).

LUNDSGAARD, E., 1939: Läkare förenings forhandlinger XLV, page 141. Quoted in ROSENBERG and WILBRANDT, 1952.

Mackler, B., and G. M. Guest, 1953: Effects of Insulin and Glucose on Utilization of Fructose by Isolated Rat Diaphragm. Proc. Soc. exper. Biol. a. Med. **83**, 327–329.

Mager, J., and M. Aschner, 1947: Biological Studies on Capsulated Yeasts. J. Bac. **53**, 283–295.

Maizels, M., 1954: Cation Transport in Erythrocytes. Symp. Soc. exper. Biol. **8** (in press).

Mandels, G. R., 1951: The Invertase of *Myrothecium Verrucaria Spores.* Amer. J. Bot. **38**, 213–221.

— 1953 a: Localization of Carbohydrases at the Surface of Fungus Spores by Acid Treatment. Exper. Cell Research **5**, 48–55.

— 1953 b: The Properties and Surface Location of an Enzyme Oxidizing Ascorbic Acid in Fungus Spores. Arch. Biochem. Biophys. **42**, 164–173.

Marnay, A., and D. Nachmansohn, 1938: Choline Esterase in Voluntary Muscle. J. Physiol. **92**, 37–47.

Mathieu, Fr., 1935: Die Resorption von Hexose-di- und -monophosphorsäure im Vergleich zu anderen Hexosen. Biochem. Z. **276**, 49–54.

McQuillen, K., 1951: The Bacterial Surface. IV. Effect of Streptomycin on the Electrophoretic Mobility of *Escherichia coli* and *Staphylococcus aureus.* Biochim. Biophys. Acta **7**, 54–60.

Merritt, H. H., 1952: Nerve Impulse. Vol. 3, Josiah Macy Jr. Found.

Meyer, K., 1954: Personal Communication.

Miles, A. A., and N. W. Pirie, 1949: The Nature of the Bacterial Surface. Blackwell Scientific Publications, Oxford.

Miller, L. L., 1954: Personal Communication.

— and W. F. Bale, 1954: Synthesis of all Plasma Protein Fractions except Globulins by the Liver. J. exper. Med. **99**, 125–132.

— and C. G. Bly, 1951: Mechanism of Plasma Protein Synthesis by the Liver as Studied with the Aid of Lysine-$\varepsilon$-$C^{14}$. Fed. Proc. **10**, 224.

— — and W. F. Bale, 1954: Plasma and Tissue Proteins Produced by Non-hepatic Rat Organs as Studied with Lysine$\varepsilon$-$C^{14}$. J. exper. Med. **99**, 133–153.

— —, M. L. Watson and W. F. Bale, 1950: Plasma Protein Synthesis Observed in Direct Study of the Liver with Aid of Lysine-$\varepsilon$-$C^{14}$. Fed. Proc. **9**, 206–207.

Mitchell, P., 1953: Transport of Phosphate Across the Surface of *Micrococcus pyogenes:* Nature of the Cell „Inorganic Phosphate". J. gen. Microbiol. **9**, 273–287.

— and J. Moyle, 1951: The Glycerophospho-protein Complex Envelope of *Micrococcus pyogenes.* J. gen. Microbiol. **5**, 981–992.

Morrison, W. L., and H. Neurath, 1953: Proteolytic Enzymes of the Formed Elements of Human Blood. I. Erythrocytes. J. biol. Chem. **200**, 39–51.

Moskowitz, M., and M. Calvin, 1952: On the Components and Structure of the Human Red Cell Membrane. Exper. Cell Research **111**, 33–46.

Mullins, L. J., 1942: Permeability of Yeast Cells to Radiophosphate. Biol. Bull. **83**, 326–333.

Myrback, K., and B. Oertenblad, 1936: Trehalose und Hefe. I. Biochem. Z. **288**, 329–337.

— — 1937: Trehalose und Hefe. II. Mitteilung: Trehalasewirkung von Hefepräparaten. Biochem. Z. **291**, 61–69.

— and E. Vasseur, 1943: Über die Lactosegärung und die Lokalisation der Enzyme in der Hefezelle. Z. Physiol. Chem. **277**, 180.

Nachmansohn, D., 1950: Nerve Impulse. Conference 1, Josiah Macy Jr. Found.

— 1951: Nerve Impulse. Conference 2, Josiah Macy Jr. Found.

— and H. B. Steinbach, 1942: Localization of Enzymes in Nerves. I. Succinic Dehydrogenase and Vitamin $B_1$. J. Neurophysiol. **5**, 109–120.

— —, A. L. Machado, and S. Spiegelman, 1943: Localization of Enzymes in Nerves. II. Respiratory Enzymes. J. Neurophysiol. **6**, 203–211.

— and I. B. Wilson, 1951: The Enzymic Hydrolysis and Synthesis of Acetylcholine. Adv. Enzymol. **12**, 259–339.

Nickerson, W. J., E. J. Krugelis, and N. Andresen, 1948: Localization of Alkaline Phosphatase in Yeast. Nature **162**, 192.

— and L. J. Mullins, 1948: Riboflavin Enhancement of Radioactive Phosphate Exchange by Yeasts. Nature **161**, 939.

OPARIN, A. I., and V. V. YURKEVICH, 1949: Adsorption of Enzymes by Yeast Cells. Doklady Akad. Nauk S. S. S. R. **66**, 247.

OSTERHOUT, W. J. V., 1952: Mechanism of Accumulation in Living Cells. J. gen. Physiol. **35**, 579–594.

PALEUS, S., 1947: On the Localization of the Specific Choline Esterase in Human Blood. Arch. Biochem. **12**, 153–154.

PARK, C. R., 1952: The Effect of Insulin and Hormones of the Pituitary and Adrenal Cortex on the Glucose Uptake by the Tissues. Vol. II, pages 634–653, A Symposium on Phosphorus Metabolism, ed. by W. D. McELROY and B. GLASS, The Johns Hopkins Press, Baltimore.

-- and L. H. JOHNSON, 1953: The Effect of Insulin on the Distribution of Free Glucose in Muscle. XIXth Internat. Physiol. Cong., 661.

PARPART, A. K., and R. BALLENTINE, 1952: Molecular Anatomy of the Red Cell Plasma Membrane. Page 135, Modern Trends in Physiology and Biochemistry, edited by BARRON, E. S. G. Academic Press, New York.

— and J. F. HOFFMAN, 1952: Acidity vs. Acetylcholine and Cation Permeability of Red Cells. Fed. Proc. **11**, 117.

PONDER, E., 1948: Hemolysis and Related Phenomena. Grune and Stratton, New York.

POPJAK, G., 1950: Mechanism of Absorption of Inorganic Phosphate from Blood by Tissue Cells. Nature **166**, 184–186.

PORTER, C. J., R. HOLMES, and B. F. CROCKER, 1953: The Mechanism of Synthesis of Enzymes. II. Further Observations with Particular Reference to the Linear Nature of the Time Course of Enzyme Formation. J. gen. Physiol. **37**, 271–289.

PRANKERD, T. A. J., and K. I. ALTMAN, 1954: Phosphate Metabolism in Normal and Pathological Mammalian Erythrocytes. Fed. Proc. **13**, 113.

PRESTON, R. D., 1952: Biological Units of Cellulose Structure. Symp. Soc. exper. Biol. **6**, 348–357.

QUASTEL, J. H., 1926: Dehydrogenations Produced by Resting Bacteria. IV. A Theory of the Mechanism of Oxidations and Reductions *in Vivo*. Biochem. J. **20**, 166–194.

— and W. R. WOOLDRIDGE, 1927 a: The Effects of Chemical and Physical Changes in Environment on Resting Bacteria. Biochem. J. **21**, 148–168.

— — 1927 b: Experiments on Bacteria in Relation to the Mechanism of Enzyme Action. Biochem. J. **21**, 1224–1251.

RAHN, O., and M. LEET, 1949: Adaptive Enzymes Induced by Insoluble Substrates. J. Bact. **58**, 714–715.

RAPOPORT, S., and J. LUEBERING, 1950: The Formation of 2,3-diphosphoglycerate in Rabbit Erythrocytes: The Existence of a Diphosphoglycerate Mutase. J. biol. Chem. **183**, 507–516.

ROBERTS, I. Z., and E. L. WOLFFE, 1951: Utilization of Labeled Fructose-6-Phosphate and Fructose-1,6-Diphosphate by *Escherichia Coli*. Arch. Biochem. Biophys. **33**, 165–166.

ROSENBERG, T., 1948: On Accumulation and Active Transport in Biological Systems. I. Thermodynamic Considerations. Acta Chem. Scand. **2**, 14–33.

— and W. WILBRANDT, 1952: Enzymic Processes in Cell Membrane Penetration. Internat. Rev. Cyt. **1**, 65–92.

ROSS, E. J., 1952: The Influence of Insulin on the Permeability of the Blood Aqueous Barrier to Glucose. J. Physiol. **116**, 414–423.

ROSS, M. H., and J. O. ELY, 1949: Alkaline Phosphatase Activity and Desoxyribonucleic Acid. J. cellul. a. comp. Physiol. **34**, 71–95.

ROTHSTEIN, A., 1954: Enzyme Systems of the Cell Surface Involved in the Uptake of Sugars by Yeast. Symp. Soc. exper. Biol. **8** (in press).

— and H. BERKE, 1952: Endogenous Alcoholic Fermentation in Yeast Induced by 2,4-Dinitrophenol. Arch. Biochem. Biophys. **36**, 195–201.

— and M. BRUCE, 1954: A Second Hexokinase in Yeast (in preparation).

-- and C. DEMIS, 1953: The Relationship of the Cell Surface to Metabolism. The Stimulation of Fermentation by Extracellular Potassium. Arch. Biochem. Biophys. **44**, 18–29.

— — 1954: The Relationship of the Cell Surface to Metabolism. XI. The Effect of Extracellular pH on Uptake of Glucose (in preparation).

— — and M. BRUCE, 1954: Fermentation of Glucose by a Cell Free Particulate Fraction of Yeast (in preparation).

Rothstein, A., A. Frenkel, and C. Larrabee, 1948: The Relationship of the Cell Surface to Metabolism. III. Certain Characteristics of the Uranium Complex with Cell Surface Groups of Yeast. J. cellul. a. comp. Physiol. 32, 261–274.
— and A. Hayes, 1954: The Relationship of the Cell Surface to Metabolism. XII. The Effects of Surface-Bound Bivalent Cations on Glucose Uptake by Yeast (in preparation).
— and C. Larrabee, 1948: The Relationship of the Cell Surface to Metabolism. II. The Cell Surface of Yeast as the Site of Inhibition of Glucose Metabolism by Uranium. J. cellul. a. comp. Physiol. 32, 247–260.
— and R. Meier, 1948: The Relationship of the Cell Surface to Metabolism. I. Phosphatases in the Cell Surface of Living Yeast Cells. J. cellul. a. comp. Physiol. 32, 77–96.
— — 1949: The Relationship of the Cell Surface to Metabolism. IV. The Role of Cell Surface Phosphatases of Yeast. J. cellul. a. comp. Physiol. 34, 97–114.
— — 1951: The Relationship of the Cell Surface to Metabolism. VI. The Chemical Nature of Uranium-Complexing Groups of the Cell Surface. J. cellul. a. comp. Physiol. 38, 245–270.
— — 1954: Unpublished Observations.
— — and L. Hurwitz, 1951: The Relationship of the Cell Surface to Metabolism. V. The Role of Uranium-Complexing Loci of Yeast in Metabolism. J. cellul. a. comp. Physiol. 37, 57–82.
— — and T. G. Scharff, 1953: The Relationship of the Cell Surface to Metabolism. IX. Digestion of Phosphorylated Compounds by Enzymes Located on Surface of Intestinal Cell. Amer. J. Physiol. 173, 41–46.
Runnström, J., 1952: The Cell Surface in Relation to Fertilization. Symp. Soc. exper. Biol. 6, 39–88.
Russell, S., 1954: The Relationship Between Metabolism and the Accumulation of Ions by Plants. Symp. Soc. exper. Biol. 8 (in press).
Sacks, J., 1949: Fractionation Procedure for the Acid-Soluble Phosphorus Compounds of Liver. J. biol. Chem. 181, 655–666.
— 1952: The Effect of Insulin on Phosphorylations in Muscle. Pages 653–664, vol. II, A Symposion on Phosphorus Metabolism, edited by W. D. McElro and B. Glass. The Johns Hopkins Press, Baltimore.
— 1953: Isotopic Tracers in Biochemistry and Physiology. McGraw-Hill, New York.
— and F. M. Sinex, 1953: Insulin and the Relation Between Phosphate Transport and Glucose Metabolism. Amer. J. Physiol. 175, 353–357.
Salton, M. R. J., 1953: Studies of the Bacterial Cell Wall. IV. Composition of the Cell Walls of Some Gram-positive and Gram-negative Bacteria. Biochem. et Biophys. Acta 10, 512–523.
Sawyer, C. H., C. Davenport, and L. M. Alexander, 1950: Sites of Cholinesterase Activity in Neuromuscular and Ganglionic Transmission. Anat. Rec. 106, 287–288.
Schmidt, G., 1951: The Biochemistry of Inorganic Pyrophosphates and Metaphosphates. Pages 443–475, vol I.. A Symposion on Phosphorus Metabolism. edited by W. McElroy and B. Glass. Johns Hopkins Press, Baltimore.
—, L. Hecht, and S. J. Thanhauser, 1949: The Effect of Potassium Ions on the Absorption of Orthophosphate and the Formation of Metaphosphate by Bakers' Yeast. J. biol. Chem. 178, 733–742.
Seaman, G. R., 1951: Localization of Acetylcholinesterase Activity in the Protozoan, Tetrahymena geleii S. Proc. Soc. exper. Biol. a. Med. 26, 169–170.
Shanes, A. M., 1952: Ionic Transfer in Nerve in Relation to Bio-electrical Phenomena. Ann. N. Y. Acad. Sci. 55, 3–36.
Slein, M. W., G. T. Cori, and C. F. Cori, 1950: A Comparative Study of Hexokinase from Yeast and Animal Tissue. J. biol. Chem. 186, 763–780.
Smith, H. W., 1951: The Kidney. Chapter V, page 81. Oxford University Press, N. Y.
Solomon, A. K., 1952: The Permeability of the Human Erythrocyte to Sodium and Potassium. J. gen. Physiol. 36, 57–110.
Sonneborn, T. M., 1951: The Role of the Genes in Cytoplasmic Inheritance. Page 291, chapter 14, in Genetics in the 20th Century, edited by Dunn. The MacMillan Co., N. Y.
Sperber, E., 1942: Aneurin und Bäckerhefe. II. Mitteilung. Biochem. Z. 313, 62–74.

SPIEGELMAN, S., M. D. KAMEN, and M. SUSSMAN, 1948: Phosphate Metabolism and the Dissociation of Anaerobic Glycolysis from Synthesis in the Presence of Sodium Azide. Arch. Biochem. 18, 409–436.

— C. C. LINDEGREN, and G. LINDEGREN, 1945: Maintenance and Increase of a Genetic Character by a Substrate-Cytoplasmic Interaction in the Absence of a Specific Gene. Proc. Nat. Acad. Sci. 31, 95–102.

— J. M. REINER, and R. COHNBERG, 1947: The Relation of Enzymatic Adaptation to the Metabolism of Endogenous and Exogenous Substrates. J. gen. Physiol. 31, 27–49.

— — and I. MORGAN, 1947: The Apoenzymatic Nature of Adaptation to Galactose Fermentation. Arch. Biochem. 13, 113–125.

STACEY, M.. 1949: The Nature of the Surface of Gram-positive Bacteria. Chapter III, in The Nature of the Bacterial Surface, edited by MILES and PIRIE. Blackwell Scientific Publications, Oxford.

STADIE, W. C., 1951: The Combination of Insulin with Tissue. Ann. N. Y. Acad. Sci. 54, 671–683.

— 1953: Studies on the Action of Insulin in Vitro. XIXth Internat. Physiol. Congress, 24–28.

— 1954: Current Concepts of the Action of Insulin. Physiol. Rev. 34, 52–100.

STEINBACH, H. B., 1951: Permeability. Ann. Rev. Physiol. 8, 21–40.

— 1954: The Regulation of Sodium and Potassium in Muscle Fibres. Symp. Soc. exper. Biol. 8 (in press).

STETTEN, D. W., 1953: Interrelationship of Carbohydrate and Fat Metabolism. XIXth Internat. Physiol. Congr. 16–18.

STEWARD, F. C., and F. K. MILLAR, 1954: Salt Accumulation in Plants: A Reconsideration of the Role of Growth and Metabolism. A. Salt Accumulation as a Cellular Phenomenon. Symp. Soc. exper. Biol. 8 (in press).

STREET, H. E., and J. S. LOWE, 1950: The Carbohydrate Nutrition of Tomato Roots. II. The Mechanism of Sucrose Absorption by Excised Roots. Ann. Bot. 14, 307–329.

THORSELL, W., and K. MYRBACK, 1951: Insoluble Saccharase in Baker's Yeast. Arkiv for Kemi, Band 3, 323–329.

USSING, H. H., 1949: Transport of Ions Across Cellular Membranes. Physiol. Rev. 29, 127–155.

— 1954: Active Transport of Inorganic Ions. Symp. Soc. exper. Biol. 8 (in press).

VASSEUR, E., 1951: Demonstration of a Jelly-Splitting Enzyme at the Surface of the Sea-Urchin Spermatazoon. Exper. Cell. Res. 2, 144–146.

VERZAR, F., and E. S. McDOUGALD, 1936: Absorption from the Intestine. Longmans Green and Co., New York.

VILLEE, C. A., V. K. WHITE, and A. B. HASTINGS, 1952: Metabolism of $C^{14}$-Labeled Glucose and Pyruvate by Rat Diaphragm Muscle in Vitro. J. Biol. Chem. 195, 287–297.

VISHNIAC, W., 1950: The Antagonism of Sodium Tripolyphosphate and Adenosine Triphosphate in Yeast. Arch. Biochem. 26, 167–172.

WALAAS, E., and O. WALAAS, 1952: Effect of Insulin on Rat Diaphragm under Anaerobic Conditions. J. biol. Chem. 195, 367–373.

WATSON, M. L., 1954: Unpublished observations.

WELSH, J. H., 1948: Symposium on the Physiology of Acetylcholine. IV. Concerning The Mode of Action of Acetylcholine. Bull. Johns Hopkins Hosp. 83, 568–579.

WERTHEIMER, E., 1934: Über die ersten Anfänge der Zuckerassimilation. Versuche an Hefezellen. Protoplasma 21, 522–560.

WESTENBRINK, H. G. K., E. P. STEYN-PARVE, and H. VELDMAN, 1947: On the Synthesis and Decomposition of Aneurin-pyrophosphate by Living Yeast. Biochem. et Biophys. Acta 2, 154–174.

WHISTLER, R. L., and C. C. SMART, 1953: Polysaccharide Chemistry. Academic Press, N. Y.

WICK, A. N., and D. R. DRURY, 1953 a: Insulin and Volume of Distribution of Galactose and Mannose. Amer. Chem. Soc. Abst. of Papers, March 15–19, 46 C.

— — 1953 b: Action of Insulin on Volume of Distribution of Galactose in the Body. Amer. J. Physiol. 173, 229–232.

— —, and E. M. MacKAY, 1951: The Disposition of Glucose by the Extrahepatic Tissues. Ann. N. Y. Acad. Sci. 54, 684–692.

Wilbrandt, W., 1954: Secretion and Transport of Non-Electrolytes.  Symp. Soc. exper. Biol. **8** (in press).
— and L. Laszt, 1933: Untersuchungen über die Ursachen der selektiven Resorption der Zucker aus dem Darm. Biochem. Z. **259**, 398.
Wilkes, B. G., and E. T. Palmer, 1932: Similarity of the Kinetics of Invertase Action *in Vivo* and *in Vitro*.  II. J. gen. Physiol. **16**, 233–242.
Wilkinson, J. F., 1949: The Pathway of the Adaptive Fermentation of Galactose by Yeast.  Biochem. J. **44**, 460–467.
Willstätter, R., and C. D. Lowry, Jr., 1925: Invertinverminderung in der Hefe. Elfte Abhandlung zur Kenntnis des Invertins. Z. Physiol. Chem. **150**, 287–355.
Young, F. G., 1953: Interrelationship of Pituitary and Pancreas.  XIXth Internat. Physiol. Congr. Abstracts of Communication, page 18.
Zamenhof, S., H. E. Alexander, and G. Leidy, 1953: Studies on the Chemistry of the Transforming Activity.  I. Resistance to Physical and Chemical Agents. J. exper. Medicine **98**, 373–397.
Zilversmit, D. B., C. Entenman, and M. C. Fishler, 1943: On the Calculation of "Turnover Time" and "Turnover Rate" from Experiments Involving the Use of Labeling Agents.  J. gen. Physiol. **26**, 325–331.

# Tension at the Cell Surface

By

## E. NEWTON HARVEY

Biology Department, Princeton University, Princeton, New Jersey, U. S. A.

With 13 Figures

## Contents

## Historical

Although phenomena connected with surface tension must have been noticed by many observers since the study of nature began, important scientific observations date from the latter part of the 17th century. The subject was considered by ROBERT BOYLE, whose contribution to the Royal Society in 1676 was entitled: "New experiments about the Superficial Figures of Fluids, especially of Liquors contiguous to other Liquors." In this paper he stated that Mr. HOOKE [1] had already enquired why water rises "in narrow pipes" and BOYLE expressed his own opinion that the concave figure of the surface (the meniscus) was determined by the contiguous fluid, the air. He endeavored to prove his point by placing another fluid, turpentine, on the water surface in a small tube, when the concavity immediately decreased and the surfaces of contact became nearly plane.

---

[1] See "Micrographia" (1665, p. 10) by ROBERT HOOKE, and his pamphlet in 1661.

He noted that a drop of one non-miscible liquid in another was spherical if the drop was small enough, but flattened if larger, and he studied the reflection of light from the boundary.  Boyle also found that under certain conditions two droplets would not mix when touched together.  Evidently a film had formed.  Boyle's experiments thus contain inquiries concerning interfacial phenomena which have occupied the attention of scientists to the present time.  The importance of the problems was realized by Boyle, who expressed his opinion in the following words:

"If these Trials and Hints, as mean as they are, be prosecuted by Naturalists that have mathematical Heads, perhaps they may conduce more to the *Physical Theory* of the Grand *System of the World,* than at first one would suspect."

No truer prophecy could have been made, especially for the living world.

The paradoxical rise of water in a small tube was studied by a number of later investigators.  Louis Carré in 1705 noted that the rise was greater the smaller the diameter of the tube and Hauksbee in 1706 that the effect was independent of the air and occurred in a vacuum.  In 1709, Francis Hauksbee, that self-taught successor to Robert Hooke as Curator of Instruments for the Royal Society, published his book, "Physico-Mechanical Experiments", containing, among many other investigations, a study of the rise of water between the surfaces of two glass plates, as well as in a capillary tube.  Hauksbee found that the rise was independent of the thickness of the glass and attributed the effect to an attraction of the glass particles nearest the liquid.

In 1718 and 1719, James Jurin published two papers in the Philosophical Transactions of the Royal Society in which he showed that in a small tube whose diameter varied the height to which a liquid would rise depended on the tube diameter at the level of the meniscus rather than at any other region.  The rise in small tubes, capillarity, thus came to be associated with surface forces, and the connection was so strong that for two centuries capillary action became a synonym for surface tension and capillary chemistry a subject dealing with chemical phenomena at surfaces.  The word, capillarity, was long a preferred term.

The attempt to account for the shape of droplets, based on the attractive forces at the surface acting through very small distances and exerting a surface tension, was made by J. A. von Segner in 1751, and firmly established by T. Young in 1804 and P. L. Laplace in 1806. Later mathematical treatment is associated with the names of C. F. Gauss and S. D. Poisson in the 1830's and G. L. van Mensbrugge and J. A. F. Plateau in the 1870's.  These investigations form the basis for many applications of surface phenomena in biology.

Knowledge of surface forces was in fact far ahead of concepts regarding the living cell.  When the rounding up of isolated fragments of protoplasm was first observed, early in the 19th century, the effect was attributed to contractility, but it was soon realized (Hofmeister 1867, Englemann 1869), that the spherical form of protoplasm was not different from that of a liquid drop and was a manifestation of surface tension rather than a vital activity. Today the importance of surface forces in determining cell shapes is universally recognized, and has been exhaustively treated in D'ARCY THOMPSON's great book, "Growth and Form" (1916, 1942).

In 1868 T. H. HUXLEY had designated protoplasm as the "physical basis of life". The time was ripe for investigation of the physical properties of the living material, particularly the forces responsible for the active movement observed in so many cells. Most of the earlier observers like HOFMEISTER (1867), thought in terms of forces within the protoplasm— a change in the water capacity of molecules in different regions of a myxomycete which led to streaming from one place to another. ENGLE-MANN's (1873) theory of muscle contraction was similar, water absorption by regions of the fibril.

However, the increasing interest in surface tension soon directed attention to the exterior surface of amebas and free-moving organisms. As early as 1855, WEBER observed with the microscope regular movements of particles when a drop of an alcoholic solution of resin is brought in contact with water, and in 1860 WRIGHT described streaming and formation of projections in a mercury drop when an electric current is passed through it. WRIGHT (1867) used these observations to explain the movements of Protozoa as follows: "It is probable that in all these cases the movements are due to modifications of the molecular attraction of the tissue, caused by a corresponding modification of the vital force, analogous to the modifications of the cohesion and molecular forces effected in inorganic matter by alteration of electrical polarity." Ten years later, GAD (1878) published his figure of a cod-liver oil drop treated with soda solution which looked exactly like a radiolarian with many pseudopods.

One of the most influential supporters of the cell surface as an important controller of cell activity was W. BERTHOLD, in whose book, "Studien über Protoplasma-Mechanik" (1886), the form of cells was attributed to surface forces and the movements of ameba and streaming protoplasts to local changes in surface tension. In the case of protoplasmic rotation in plant cells, the movement was believed to be due to differences in surface tension in the limiting layer of the cell sap and the endoplasm.

Perhaps the most striking attempts to imitate cell movements by means of models was made by G. H. QUINCKE (1888), long interested in capillary phenomena at interfaces (1870, 1877). Using a mixture of almond oil and chloroform, which sank in a dish of water to a flattened circular drop, QUINCKE found that projections would be formed and movements occur, similar to those of amebas, if 2% sodium carbonate solution were brought near the drop by a capillary tube. The observations led to his view that the protoplasm might be surrounded by a film of oil, less than 0.0001 mm thick. Amoeboid movements would occur through the action of albumen on the inner surface of the oil, producing a substance (according to QUINCKE an albuminous soap) which spreads, just as the alkali produces a fatty acid soap, spreading over the almond oil surface, lowering its surface tension, and resulting in movement.

Thus was born the oil theory of the nature of the protoplasmic surface, a view which gained many adherents when studies on the permeability of cells by E. Overton in the years 1895–1900 indicated the rapidity with which oil soluble substances can penetrate. Overton spoke of a "lipoid film" at the surface, considering lecithin to be the principal lipoid involved, and later workers substituted the concept of a surface made up of a mosaic of protein and cholesterol or protein and lecithin in order to account for the penetration of non-lipoid soluble substances.

The experiments of Quincke stimulated Otto Bütschli to continue the studies, largely because of his emulsion theory of the structure of protoplasm. Bütschli (1889) found that by mixing olive oil with finely ground potassium carbonate and placing the mixture in water, an artificial ameba could be produced which moved about by pseudopod formation for several days in a surprising imitation of the living cell. The carbonate forms minute vacuoles of soap which burst on reaching the surface, lowering the surface tension of the oil locally, with bulging at this point, followed by streaming movements. These views were summarized in Bütschli's book, "Untersuchungen über mikroskopische Schäume und das Protoplasma" (Leipzig, 1892; English translation 1894), and no doubt laid the background for the tacitly accepted belief that the value for surface tension of a naked cell must be approximately like the interfacial tension between oil and water. Certainly the experiments on oil-water movement exerted a profound influence on the explanation of amoeboid movement for many years. Only twenty-five years ago, Krizenecky and Dubska (1927) published "Eine Methode zur Messung der Oberflächenspannung biologischer Flüssigkeiten gegen ein Protoplasma-ähnliches Medium" in *Protoplasma*. The protoplasmic medium was paraffin oil and the method that of an adhering ring, pulled away from the oil interface in aqueous solution.

It is interesting to note that Quincke, Berthold and Bütschli did not speculate on the actual value of the tension at the surface of cells, and that no attempt to determine the tension is mentioned in their work. In the writings of other investigators of protoplasm who adopted the surface tension theory of movement, such as Verworn (1892, 1895), Rhumbler (1898, 1905, 1914), or Jensen (1901, 1902), there are also no values for tension. Hirschfeld (1909) and McClendon (1911) looked on amoeboid movement as due to surface tension changes connected with variation in interfacial electric potential and K. Gruber (1912), Tait (1920), Haberlandt (1919) and Fürth (1922) also adopted the surface tension theory of movement, again without mention of a value.

The possible importance of surface tension was also recognized in muscle physiology. Bernstein (1900), who had studied shape changes of a mercury drop in contact with a crystal of bichromate[2], movements resulting from local lowering of surface tension, based his theory of muscle

---

[2] Bernstein mentioned Paalzow (Pogg. Ann. d. Physik **104**, 419, 1858) as the first to observe oscillatory movements of mercury in acid solution in contact with bichromate.

contraction on surface tension changes in contractile particles. However, there were many (e. g. EWART 1903, JENNINGS 1906) who did not accept the surface tension theories. TIEGS (1928) has fully reviewed "Surface tension and the theory of protoplasmic movement", as applied to muscle, amoeboid changes and protoplasmic streaming, while DE BRUYN (1947) has dealt exhaustively with "Theories of amoeboid movement". Suffice it to say that sol-gel changes are now regarded as far more important in amoeboid movement and protoplasmic streaming than changes in surface tension (see MARSLAND, 1942).

Another vital activity which has been attributed to surface tension change is the division of an animal cell into two. Streaming movements like those accompanying amoeboid movement, have long been observed in dividing cells[3]. In SPEK's (1918) review of the subject, "Oberflächenspannungsdifferenzen als eine Ursache der Zellteilung", the first research on mechanics of cell division is attributed to BÜTSCHLI (1876, 1900), who regarded an increase in surface tension at the equator subsequent to nuclear division as the principal factor in cutting a cell into two. RHUMBLER (1896, 1899) was particularly impressed by the currents in dividing cells.

R. S. LILLIE (1903) also brought his view of the distribution of electric charges in cells during cell division in line with a theory of cleavage connected with differences in surface tension. On the LIPPMANN-HELMHOLTZ principle that the greater the electric potential difference across a surface, the less the surface tension, LILLIE came to the conclusion that there was a greater lowering of surface tension at the poles than at the equator. Later LILLIE (1909) changed his viewpoint to that of ROBERTSON (1909, 1911) who claimed, on the basis of oil-soap models that a region of lowered surface tension at the equator would divide a cell. McCLENDON (1910, 1912) criticized the model, holding that relatively higher tensions must be present at the equator, if surface tension is to be effective in cell division. The controversy is chiefly of historical interest, since theories of cleavage, like those of amoeboid movement, now invoke sol-gel changes at the cleavage furrow as the primary cause of fission, rather than surface tension (MARSLAND, 1951).

Theories have also been proposed in which the effect of a substance on the activity of the cell has been attributed to a change in surface tension. For example, HEILBRUNN (1913, 1915, 1924, 1925) has held that the formation of a fertilization membrane on unfertilized sea urchin eggs is due to a lowering of surface tension, since he found that the majority of parthenogenetic agents lower the surface tension of the egg surface. When the surface tension is lowered sufficiently, cytolysis results (1915).

It will thus be observed that the tension at the surface of cells has been invoked as a primary factor in various biological functions by a wide variety of investigators, but the magnitude of the forces involved has been

---

[3] See the excellent discussion in WILSON, "The Cell in Development and Heredity", 3rd Ed., New York, pp. 192–197, 1925.

mostly neglected. PFEFFER (1891) studied the cohesive force of a slime-mold (*Chondrioderma*) thread of 0.3 mm diameter, which could just support a weight in air of 3.5 mg for 1½ min. If we assume the force is all due to the tension around the circumference of the thread the value is 37 dynes/cm. PFEFFER also quoted QUINCKE (1888) as having found an inter-facial tension between fresh olive oil and concentrated egg white solution of 3.3 dynes/cm.

R. HOEBER assumed a value of 0.01 gram/cm (9.8 dynes/cm) for proto-plasmic tension in the first edition of "Physikalische Chemie der Zelle und Gewebe" (1902), but in the second edition (1906) he calculated another probable value. Taking the surface tension $\gamma$ of air-water as 0.082 grams per cm and subtracting from this the surface tension of an air-egg albumen solution (0.059 grams per cm), he arrived at the surface tension of the protoplasm-water interface as 0.023 grams per cm (22.5 dynes/cm). This figure was used to calculate the internal pressure, $P$, of small cells, from the relation, $P = 2\gamma/r$, where $r$ is the radius of the protoplasmic sphere. The same calculation appears in the 3rd (1911), 4th (1914), and 5th (1922—1924) and 6th (1926) editions [4]. In the 5th edition (p. 386) there is an additional statement that the surface tension of protoplasm-water is like that of a sodium glycocholate solution-petroleum interface.

Another approach was made by F. CZAPEK (1910), who had noted that exosmosis of material took place when plant cells were placed in critical concentrations of a wide variety of organic compounds dissolved in water. The surface tension against air of these critical concentrations was found to be 68 to 69 if the air-water surface tension was taken as 100, and CZAPEK supposed that the protoplasm-water interfacial tension must therefore be approximately 0.68 of that between water and air (73 dynes/cm) or about 50 dynes/cm. There is, however, no reason why this relation should be true. Since CZAPEK's time a number of direct measurements of tension at the cell surface have been made. These will be described in detail in a later section, together with the values obtained for various cells.

## Forces and Structures at the Cell Surface

Like most problems in biology, measurement of the tensional forces at the surface of living cells presents a complex problem, which cannot be completely solved by application of simple physical concepts. If the living cell were merely an insoluble liquid, such as paraffin oil, suspended in water, it would be possible to speak of the surface tension between the two phases, *i.e.* the oil-water interfacial tension. Although isolated cells are "insoluble" in their suspended medium, this immisceability comes from the presence of a definite membrane (the plasma membrane). The interior contents of the cell are quite soluble in aqueous solutions but they exhibit the ability to gel and it is sometimes impossible to state whether a deform-

---

[4] The English sequel, "Physical Chemistry of Cells and Tissues" (Philadelphia 1945), gives modern values for tension at the surface of cells.

ing force is acting against the tension of a surface or the resistance of a gel. A peculiar layer near the surface of cells, the ectoplasm or cortex, was very early recognized as distinct from the endoplasm. The gel-like consistency of ectoplasm is easily observed in many cells, and changes in the rigidity of this layer have been measured by W. L. WILSON (1951) and WILSON and HEILBRUNN (1952).

It is probably not correct to speak of the surface tension of the plasma membrane because the most accurate studies indicate that the surface layers of naked cells exhibit elastic properties. The tension increases as the area of the cell increases, whereas surface tension is independent of surface area. Therefore the expression "tension at the surface" or "surface forces" of living cells is less ambiguous than surface tension.

It is obvious that all gradations may exist between films one molecule thick and structures visible with the light microscope. If monolayers or multilayers of a substance are present at a surface, reproducible tensions may be measured, as in a soap film, but the value will depend on the concentration of surface-active substances. Finally the thickness of a third substance may reach the point where a thin film may be said to exist, and a third three-dimensional phase, however thin, is to be considered. A true membrane is usually solid and work must be done to bend it. Evidence of its existence is frequently obtained by crushing or tearing cells, when the frayed ends of the membrane remain as rigid solid films. However, it must not be forgotten that many monolayers behave like solids and are classified as solid films, and that liquid films may exhibit elastic properties.

The difference in behavior of a surface with elastic properties is well seen in the studies of HARVEY and DANIELLI (1936) on albumen bubbles. If the surface force, $T$, of such a bubble is measured by the internal pressure ($P$) method, $T = P r/4$, the tension, $T$, is found to increase as the radius, $r$, of the bubble increases, *i. e.* as the surface area increases. On the other hand soap bubbles exhibit no such behavior and the same is true for lecithin bubbles. Only when protein (egg white) is present does the film show elastic properties. The albumen films and mixtures of albumen and soap films also exhibit hysteresis, *i. e.* the curve for surface tension vs. area follows a different course when successive measurements are made as the bubble decreases in size when compared with measurements made as the bubble increases in size. The effect is shown in Fig. 1.

The words "surface tension" applied to living cells have always been widely used for "tension at the surface" and we have seen that surface forces have been invoked as a determining factor in such phenomena as protoplasmic streaming, cyclosis, amoeboid movement, and cell division. Since the shape of cells has long been attributed to surface tension, it is important to point out that the shape of any liquid body is practically the same, under the various imposed conditions, whether the surface is a thin elastic membrane or the mere boundary of a liquid phase. Moreover, whatever part tension at the surface may play in cyclosis, amoeboid

movement or in cell division (actual cleavage), these processes must always
work against the sum of surface forces which tend to prevent any increase
in surface area, be they pure surface tension, elastic films or a gelled
network.

In this account, only those cells will be considered whose surface is
obviously liquid in behavior, *i. e.* is readily distorted and reconstitutes itself when broken without leaving frayed edges. Whether a "pellicle" is
present or not, a determination of the tension and of the behavior on
stretching will give valuable information on the order of magnitude of
surface forces to be reckoned with in the living cell. Such cells can readily

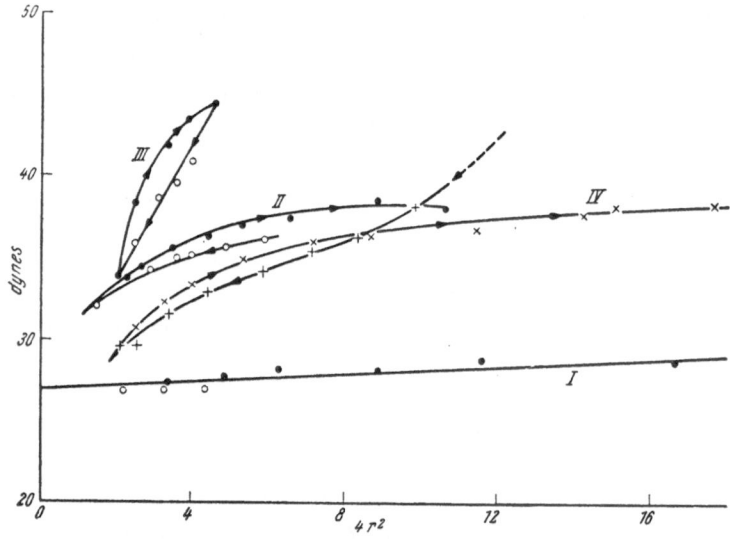

Fig. 1. Change in surface tension (dynes/cm) as a function of bubble area (4 $r^2$). I. Soap solution; II. Soap solution 1 part, egg white 4 parts;  III. Egg white;  IV. Soap solution 1 part, egg white 2 parts. The arrows indicate the order in which measurements were taken.
After Harvey and Danielli (1936).

increase their surface area. They undergo movement (Amoebae, leucocytes,
slime molds), or can readily be fragmented by shaking or centrifuging
(certain marine eggs), or they may be protoplasts, in which the obvious
external cellulose membrane can be separated either by plasmolysis or by
microdissection.

A membrane may be semi-solid, a liquid of such high plasticity that it
retains its form against ordinary disturbing forces, but moves under greater
ones. Such appears to be the condition of the "pellicle" or vitelline membrane of unfertilized *Arbacia* and *Asterias* eggs, which can be moved by
centrifugal force in the centrifugal direction (E. B. Harvey 1932).

That the surface of an unfertilized sea urchin egg is under no great
tension can be inferred from the fact that newly laid eggs are often ir-
regular in shape and become spherical only some time after standing in
sea water, or on insemination. The surface of *Arbacia* eggs may also
become wrinkled or amoeboid. E. B. Harvey has observed such wrinkling

during a stage in development of various sea urchin eggs (1935) and amoeboid shapes after treatment of unfertilized *Arbacia* eggs with urethane (unpublished observations).

On fertilization of a sea urchin egg, the "pellicle" or vitelline membrane is elevated and hardens to form the fertilization membrane, a rigid structure. In long centrifuged *Asterias* eggs after fertilization a fertilization membrane appears only at the heavy or centrifugal end (CostELLO 1935). According to CHAMBERS and KOPAC (1937) the vitelline membrane of *Arbacia* eggs can be removed by repeated washing with isotonic NaCl solution, and urea (MOORE 1930) has been used for the same purpose. No fertilization membrane is formed after such treatments. A discussion of the surface structures of sea urchin eggs will be found in papers by CHAMBERS (1940), RUNNSTRÖM and MONNÉ (1945), RUNNSTRÖM, MONNÉ and WICKLUND (1946), and RUNNSTRÖM (1952). It would be useless to quibble over the terms film, pellicle, membrane, etc. and it is sufficient for our purpose to remember that what is measured as the tension at the surface of cells is the sum of the surface and elastic tensions of a definite molecular structure.

The tension of visible membranes may be very great. The cellulose walls of plant cells can withstand turgor pressures of many atmospheres. KAO, CHAMBERS and CHAMBERS (1951) have measured a pressure of 150 mm Hg within the chorion of eggs of the fish, *Fundulus*, in sea water, after certain "platelets" at the egg surface have dissolved. In many eggs (for example, *Nereis*) the enveloping membrane prevents breaking into two halves when the eggs are centrifuged. Such cells are obviously unsuitable for study of tension at protoplasmic surfaces.

## Methods of Measurement and Values

The methods of measuring tension at the surface of cells are all modifications of well known procedures for surface tension determination. These standard measurements may be classified in two categories: (1) Dynamic methods, in which the surface undergoes rapid movement, and (2) Static methods in which the surface is at rest or nearly at rest. The dynamic methods do not give true values for solutions, since the surface is continuously disturbed and time is not sufficient for surface active substances to collect in the surface. They may be used only for a pure liquid. Some of the static methods involve slight movement of the surface and should always be carried out in such a way as to reduce this movement to a minimum. Although by no means a complete list, the following methods may be recognized:

Dynamic.

(1) Surface waves.

(2) Vibrating jets.

(3) Vibrating drops.

Static.

(4) Capillary height.

(5) Adhesion of rings or plates.

(6) Drop weight.

(7) Internal pressure.

(8) Curvature of surfaces.

    (a) Meniscus.

    (b) Sessile drop.

    (c) Pendant drop.

The various procedures are so well known that they will not be discussed in detail. The reader is referred to standard works on surface phenomena or to such papers as those of Dorsey (1929), Kopaczewski (1933), and Herčik (1934). Inspection of the list of methods indicates that many could not be applied to a cell without serious damage. Actually nos. 3, 6, 7 and 8b have been tested with living cells, often with considerable modification, and some have yielded satisfactory results.

Vibrating drop.—The tension at the surface of cells can hardly be measured by the surface wave or vibrating jet methods but it might be possible to study the surface forces if a cell could be made to vibrate after deformation, like a drop of water falling from an orifice. The simplified equation is:

$$\gamma = 3 \pi W/8 \theta^2$$

where $W$ is weight, $\theta$ is the period of vibration and $\gamma$ is surface tension.

To test this possibility, unfertilized eggs (*Arbacia punctulata*) were forced through a glass capillary tube somewhat smaller than the diameter of the egg, after which they emerge as short cylinders with rounded ends (Harvey and Shapiro 1941). Such forms should oscillate under the influence of surface tension, rounding up and again elongating a number of times. However, it was observed that if the capillary was too small, the eggs became injured and gelled, retaining the elongated cylinder shape indefinitely. If only slightly compressed the eggs nearly rounded up in approximately 0.5 second (see Fig. 2) but never oscillated. An equation was presented for the recovery time or relaxation time, relating time, true viscosity, true surface tension and the major and minor axes of the deformed egg. However, the conditions in the egg are too complicated for any simple relation to hold.

Vlès (1933), by deforming eggs sucked into a capillary pipette (usually about 38 $\mu$ in diameter) has found even longer relaxation times, measured in minutes for the egg of the sea urchin, *Paracentrotus lividus* (about 100 $\mu$ in diameter). He also determined for the whole egg, the modulus of longitudinal deformation and transverse contraction, equivalent to Young's and Poisson's modulus, respectively.

Shapiro (1941) also studied the return to a spherical shape after an *Arbacia* egg has been elongated by centrifugal force. The conditions for recovery are much more complicated in this case because of the distribution of granules within the egg. The times involved are also measured in minutes rather than fractions of a minute as when the uncentrifuged *Arbacia* egg is used. It was found that presence of Ca retarded the recovery.

Thus, all attempts to measure tension by treating sea urchin eggs as vibrating drops have failed. Since the viscosity of the egg as a whole is

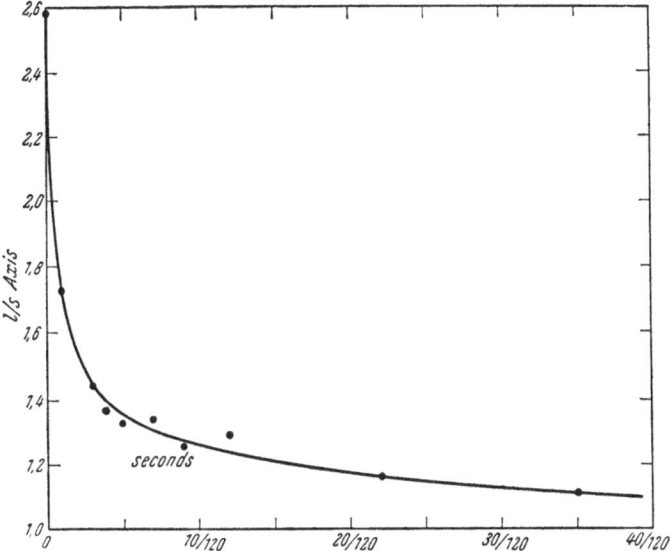

Fig. 2. The ratio of long to short axis of an unfertilized *Arbacia punctulata* egg (74 μ diameter) which has been distorted by passing through a capillary tube (50 μ in diameter) plotted against time (in ¹/₁₂₀ seconds) as rounding up occurs.
After HARVEY and SHAPIRO (1941).

probably no more than ten times that of water (see HEILBRUNN 1952), the fact that egg cells do not vibrate when deformed can be taken as indicating a low tension at the surface.

### Centrifuge method

This approach is essentially a modification of the drop weight method in which the pull of gravity, $g$, on a drop (its weight) balances the surface tension, $\gamma$, at the circumference, $2\pi r$, of the tube on which the drop hangs:

$$\text{Weight} = \varrho\,Vg = 2\,\pi\,r\gamma$$

where $\varrho$ is density, $V$ volume and $g$ the force of gravity, 980. The living cell cannot be used as a drop of liquid, but the tension around the circumference of a cell can be equated to centrifugal forces which pull it into two halves. The cell is assumed to behave like a liquid, which becomes unstable when drawn into a cylinder whose length is $\pi$ times its diameter, breaking into two or more spheres. The method is applicable to spherical cells (marine eggs) containing oil globules lighter than the cell

fluid and yolk granules heavier than the cell fluid.  Under the influence
of centrifugal force the oil and yolk separate and exert forces that pull
the cell into two almost equal spheres (Harvey 1931 b).  At the moment of
instability we may equate these forces to tension $T$ around the circumference of the elongated cylinder, thus:

$$\pi D T = C g \left[ V_h (\varrho_h - \varrho_m) + V_L (\varrho_m - \varrho_L) \right]$$

where $D =$ diameter of cylinder, $C =$ centrifugal force in times gravity,
$g = 980$,  $V_h =$ volume  of  heavy  half,  $\varrho_h =$ density  of  heavy  half,
$\varrho_m =$ density of medium, $V_L =$ volume of light half and $\varrho_L =$ density of
light half.

The whole process of separation into two half cells can be watched
(or photographed) in the centrifuge-microscope (Harvey 1932), as illustrated
in Fig. 3.  The time for breaking into two is determined by the centrifugal

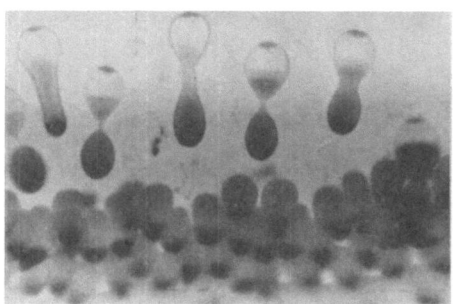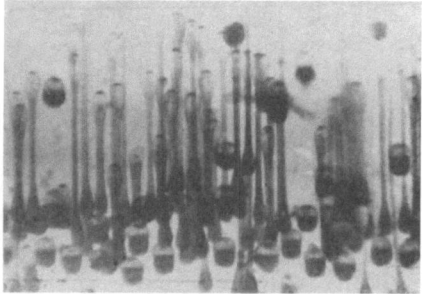

Fig. 3.  Left: Centrifuged unfertilized *Arbacia punctulata* eggs on a centrifuge-microscope slide showing
various stages in breaking apart of the egg into two halves.  Centrifugal direction down.  The order of
stratification of egg granules is oil (top), nucleus, clear layer, mitochondria, yolk, pigment (bottom).  Many
heavy half eggs are shown at bottom of slide   Right: Fertilized *Arbacia punctulata* eggs photographed in
the centrifuge-microscope, showing the long membranous stalk connecting the light and heavy ends of the egg.
After Ethel Browne Harvey.

force and the viscosity of the cell.  Therefore the force is taken as that
just sufficient to separate the two halves with infinite time.  Under these
conditions the value obtained for the tension is a maximum one.  Results
for the unfertilized eggs of a sea urchin, *Arbacia punctulata*, range around
0.2 dyne/cm. The surface of such an unstable cylinder would be about 25%
greater than the surface of the original sphere so that the value 0.2 dyne/cm
represents a 25% increase of the surface.  Such low tensions suggest that
no appreciable forces at the surface of the protoplasm oppose osmotic
swelling, and in fact a study of osmotic volume increase of the egg of the
annelid, *Ceratocephale osavai*, by Kamada and Yamamoto (1931), indicate
that the elastic constant of the plasma membrane of this egg is negligibly
small.

A modification of the centrifuge method can be applied to cells into
which oil drops have been injected.  In this case the oil drop is pulled out
by centrifugal force, and the figure becomes unstable when a neck of
protoplasm forms of diameter equal to the diameter of the oil drop.  The
buoyant force of the oil when centrifuged is equated to the tension around

the circumference of the neck of the protoplasm formed when it pulls away, thus:
$$\pi\,D\,T = C\,g\,[V_0\,(\varrho_m - \varrho_0)]$$

where $D$ = diameter of neck, $V^0$ = volume of oil drop, $\varrho_0$ = density of oil, $\varrho_m$ = density of medium surrounding cell, $Cg$ = centrifugal force.

Values obtained by this method for *Amoeba dubia* are 1–3 dynes/cm (HARVEY and MARSLAND 1932); for a slime-mold, *Physarum polycephalum*, around 0.45 dyne/cm (VEXLER 1935); for rabbit macrophages 2 dynes/cm,

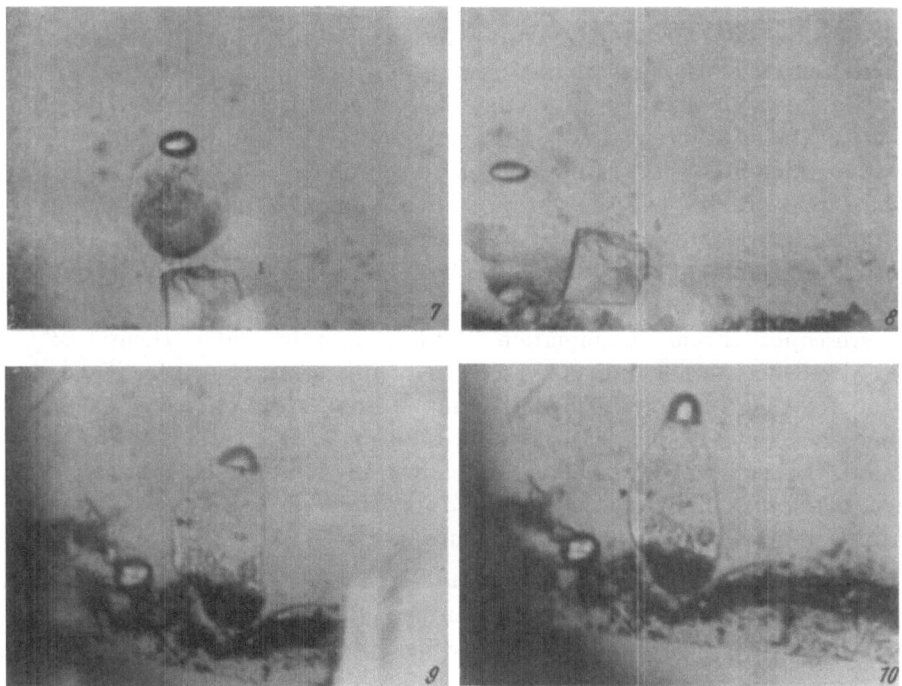

Fig. 4. *Amoeba proteus* (above) and *Amoeba dubia* (below) containing injected oil drops, photographed while revolving in the centrifuge-microscope. Note firm surface of *A. proteus*, against which the oil flattens, and the pulling out of *A. dubia* by the buoyant oil. With higher centrifugal forces the oil will pull away from *A. dubia* completely. The force in 7 is 5300; in 8, 12500; in 9, 4800; in 10, 5750 times gravity.
After HARVEY and MARSLAND (1932).

and for frog leucocytes 1.3 dynes/cm (SHAPIRO and HARVEY 1936) at room temperature (22° C). Photographs of oil in *Amoeba proteus* (above) and *Amoeba dubia* (below), taken with the centrifuge-microscope, are shown in Fig. 4.

In the drop weight ($W$) method of measuring surface tension, $\gamma$, a correction factor $f$ should be applied, because the circumference at the surface of the stalk connecting the drop with the tube at the moment of breaking is less than the circumference of the tube itself:

$$W = 2\,\pi\,r\gamma f$$

The correction factor $f$ is a complicated function of the tube diameter and the capillary constant, with a minimum of 0.6. Applying a correction factor of 0.6 to the centrifuge method equations, the above values for tension at various cell surfaces become somewhat greater.

However, it must be pointed out that the centrifuge method indicates the general magnitude of cell tensions rather than giving highly accurate absolute results. It is best adapted for comparative studies. E. B. Harvey (personal communication) has observed that unfertilized eggs from different individuals of *Arbacia* vary greatly in the centrifugal force necessary to pull them into two halves. They are also more difficult to break after long standing in sea water or late in the season.

Fertilized sea urchin eggs from which the fertilization membranes have been removed also vary in the ease with which they can be pulled into two at any one stage of development, and fertilized eggs from the same batch vary at different times of development. The results are complicated by the fact that the hyaline plasma membrane ($2-3\,\mu$ thick) develops at the surface of the egg soon after fertilization. At the time of aster and spindle figure formation gelled regions also appear in the egg. The fertilized egg is no longer as simple a physical system as the unfertilized egg. E. B. Harvey (1933) has studied the breaking process in the centrifuge microscope. From insemination until 5 minutes later (temp. $23^0$ C), fertilized eggs break into halves more readily than the unfertilized, but from 5 until 25 minutes after insemination the surface has greatly changed in consistency. A long viscous semi-solid stalk or streamer forms which holds the two halves together, as shown in Fig. 3. This stalk slowly lengthens and finally breaks, retaining its threadlike form, and making the application of the above equation meaningless.

When unfertilized *Arbacia punctulata* eggs are centrifuged in various pure isosmotic salt solutions, E. B. Harvey (1945) has found that they break into halves most readily in the following order: $KCl > NaCl >$ sea water $> MgCl_2 > CaCl_2$. The stratification of the granules in the eggs (a measure of viscosity) occurs in the reverse order, those in $CaCl_2$, which break with greatest difficulty, are most readily stratified, and hence least viscous. This reverse relation indicates that the salt action does actually change the surface, whose tension is acting against the forces drawing the egg into two halves.

The centrifugal method has also been used by Raven (1945) and de Vries (1947) for comparison of the tension at the surface of fertilized eggs of the fresh water snail, *Limnaea stagnalis*, at various stages of development, although no values for tension were obtained. The eggs do not break into two with the forces used (around $1000 \times g$), but the "tension index" is expressed by the ratio of greatest diameter at right angles to the centrifugal force to the greatest diameter in the direction of the centrifugal force. The tension index changes during development (Raven) and also in the presence of $LiCl_2$ and $CaCl_2$. De Vries found the tension to be increased in 0.6 to 1% $LiCl_2$ and 1.5% $CaCl_2$ but decreased in 0.1% $LiCl_2$ solution. Viscosity was also studied by the movement of granules during centrifuging.

## Pressure or kinetic method

A possible method of measuring the tension $T$ of a spherical cell of radius $r$ involves the simple equation, $P = 2T/r$, or $P = 4T/r$, if the cell membrane has two surfaces, like a soap bubble. It would be necessary to insert a micropipette into the cell, make sure no leakage occurred and measure the pressure, $P$. The pressures would be small and technical difficulties have generally prevented its use.

However, an order of magnitude for surface forces can be calculated from a method which depends on determining the pressure within the cell indirectly, by what might be termed a kinetic method. Just before the completion of first cleavage of an *Arbacia* egg the two blastomeres are connected by a small stalk. If one blastomere is punctured, the remaining one will discharge its contents through the stalk, due to an excess internal pressure from the tension at the surface. From moving pictures the rate of discharge can be determined by measuring the decrease in volume of the blastomere. It follows a law which would indicate elastic forces at the surface. Assuming POISEUILLE's Law for flow of liquid in a tube and using a value for viscosity previously found for the egg fluid at this stage of development, SICHEL and BURTON (1936) calculated the excess internal pressure to be 64 dynes/cm² and the tension 0.09 dyne/cm, agreeing well with values obtained by other methods.

A similar calculation by HEILBRUNN (1952, p. 372), based on the flow of protoplasm in a pseudopod of *Amoeba proteus* has indicated a pressure of 8 dynes per cm². If the pressure is due to surface tension of a sphere $30\,\mu$ in radius, the tension would be 0.012 dyne/cm, again a low value.

## Compression method

The most accurate of the pressure methods involves measuring the force $F$ necessary to flatten a spherical cell a given amount. Since $F/A = P$, and in this case the cell has two radii of curvature, $r_1$ and $r_2$, the general equation for surface force $T$:

$$F/A = P = T\left(\frac{1}{r_1} + \frac{1}{r_2}\right)$$

can be used. Again it must be assumed that the cell surface has no rigidity, *i. e.* that no force is necessary to bend the surface.

With the above equation, COLE (1932) has not only determined the tension of the unflattened *Arbacia* egg but obtained values for the increase in tension on compression, $Z$, *i. e.* as the surface area increases on flattening. The eggs are flattened by a microbeam of gold, $6\,\mu$ thick and $180\,\mu$ wide, pressing on the top of the egg, and then photographed. Fig. 5 presents a diagram of the egg outline. The two radii of curvature ($r_1$, $r_2$) can be measured from the photographs and the pressure ($P$) calculated from $F/A$, where $A$ is the area of the egg flattened by contact with the beam surface. The beam, held in such a manner that it always remains parallel to the

plane of compression, is calibrated by hanging microweights on one end
and noting its deflexion by light reflected from the beam to a scale.   The
forces to be measured are of the
order of $10^{-6}$ grams, 2 $\mu$g giving
a compression of 25 $\mu$.  The cells
behave as if they do not adhere
to the surfaces.

Results on the unfertilized egg
of *Arbacia punctulata*, 74 $\mu$ in
diameter,   give   a   tension   of
0.133 dyne/cm when compressed
25 $\mu$, and lower values for less
compression.    Extrapolation  of
the   tension-compression   curve,
which  is  slightly  convex  to  the

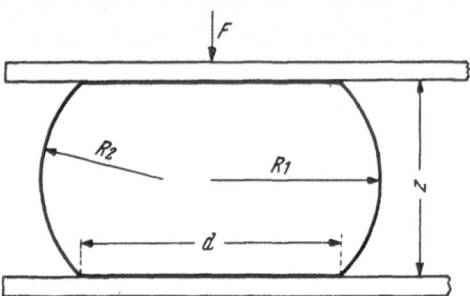

Fig. 5.   Diagram of flattened sea urchin egg showing
measurements to be made to determine tension by
compression method.
After Cole (1932).

compression axis, to zero compression, gives a value of 0.08 dyne/cm for
the uncompressed egg.   The evidence for elasticity of the surface is clear
and conclusive, as indicated
in Fig. 6.

The fertilized egg of *Arbacia* without a fertilization
membrane behaves as the unfertilized and gives the same
low values for surface forces
until shortly before first cleavage, but the fertilized egg
with fertilization membrane
behaves as if the membrane
were   decidedly   rigid   (Cole
and Michaelis 1932).   For a
fertilized egg without a membrane, the computed force for
a 5 $\mu$ compression is 0.3 dy-
ne/cm as compared with 1 dy-
ne/cm   with   a   fertilization
membrane,   but   this   latter
figure merely represents an
expression of the experimental data rather than the actual surface force, since the
effect of rigidity has not been
considered in the equation.

Another   method   of   compression is to suck whole cells
under negative pressure into
capillaries   of   considerably
smaller   diameter   than   the

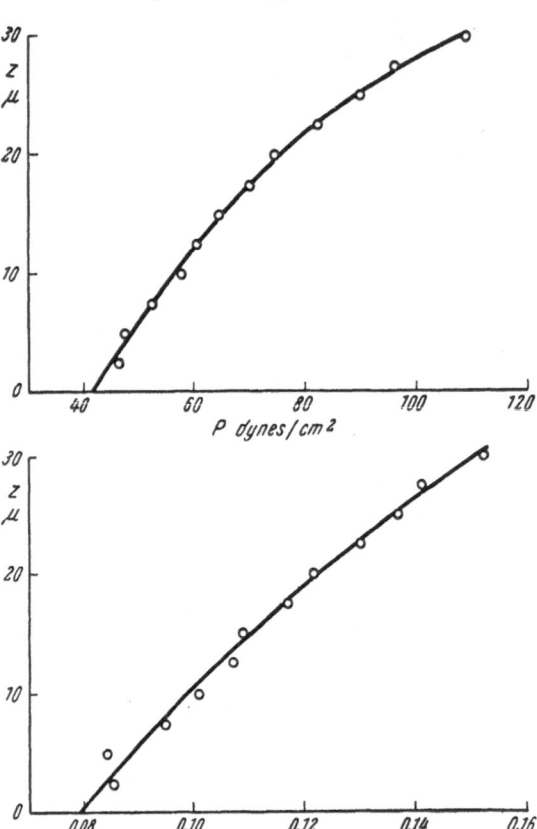

Fig. 6.   *Above:* Pressure (*P* in dynes/cm²) in an unfertilized
*Arbacia punctulata* egg as a function of compression (*Z* in
micra) between two plates.   *Below:* Surface force (*T* in
dynes/cm) for the same egg vs. compression (*Z* in micra).
Note extrapolation to zero compression indicates a tension
for the undistorted egg of 0.08 dynes/cm.

spherical cell itself. VLÈS (1933) has used this method to study the behavior of sea urchin eggs under various conditions, and, by measuring the pressure necessary, obtained values for the elasticity of protoplasm. PFEIFFER (1936a), by reasoning similar to that of HATSCHEK (1910), who studied the pressure necessary to force emulsified oil drops through small pores, developed the following equation relating the pressure $(P)$ necessary to suck a liquid sphere of radius $r$ into a capillary of smaller radius against its surface tension $(T)$:

$$T = g P r \, (a - 1)/K$$

where $K$ is a constant varying from 2.3 to 2.15, $g = 980$, and $a$ is the ratio of sphere diameter to capillary diameter. The cells used were free proto-plasts of the pulp of fruits. The values for $T$ were about 20 dynes/cm. The protoplasts are in reality thin films of protoplasm surrounding a water-filled vacuole and possess both a plasma membrane and a vacuolar membrane. Caution is necessary in applying simple relations to structures of this type. PFEIFFER (1935, 1936 b) has also studied the rate at which these protoplasts are sucked into capillaries as a function of sucking pressure and finds the relation non-linear, thus indicating plasticity. A study of the elastic properties (YOUNG's modulus) of these protoplasmic bubbles has also been made (PFEIFFER 1936 c, 1937), part of a series of investigations on quantitative determination of the molecular forces of protoplasm, published in *Protoplasma*, 1932–1937.

## Sucking method

The experiments of VLÈS (1933) in which whole sea urchin eggs are sucked into a capillary pipette in order to determine limits of rupture, permanent deformation, relaxation time, and the modulus of rigidity have already been referred to. If only a small volume of the egg is sucked into the capillary, resulting in a slight bulge at the surface, another pressure method of studying forces at the surface of cells becomes available. The technique has been developed by MITCHISON (1952) and SWANN (1952) in connection with their "expanding membrane" theory of cell division and has not been published, but they have kindly allowed me to outline the method. Measurements of the deflection of the bulge (up to the point where the bulge becomes a hemisphere) are made with a microscope, while the negative pressure of a reservoir of water connected with the pipette and raised or lowered by a micrometer screw, is read. When deflection is plotted vs. pressure (about 3 mm water is necessary for an unfertilized egg), a straight line is obtained whose slope is proportional to the elastic modulus of the membrane. The mathematical theory of the problem is not completely solved, and values for elasticity require a knowledge of the thickness of the membrane, but changes in elasticity can be detected during development of sea urchin eggs, as well as changes due to various environmental factors and poisons. The area of the cell surface need be increased only 6% and readings may be taken rapidly—with practice an egg cell every two minutes. There is every indication that the "sucker" or "cell elastimeter" will become a simple tool of considerable value.

## Stretching method

It should be possible to measure the tension of a cell surface by stretching the cell. The equation for this method is:

$$T = (F - A P)/C$$

where $T$ is the tension in dynes/cm, $F$ is the stretching force, $A$ the cross sectional area of the stretched cell at the widest point, $P$ the internally directed pressure and $C$ the circumference at the widest cross section. All the quantities are easily measured except $P$. With considerable elongation, $P$ becomes small and can be neglected, a procedure which gives tensions higher than if $P$ is known.

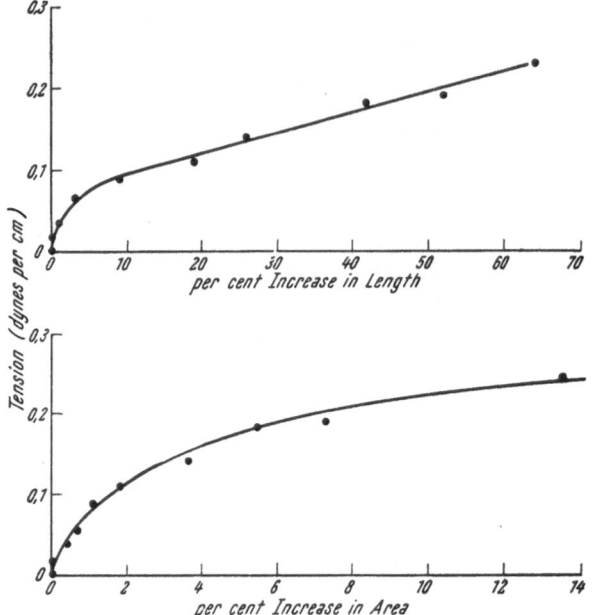

Fig. 7.   Tension at the surface of an unfertilized *Arbacia punctulata* egg as it is stretched between two glass needles. After Norris (1939).

The method has been perfected by Norris (1939). In practice the cell is impaled on needles, one of which is rigid and the other is moved, thus exerting the stretching force. The bending of the movable needle can be measured and by calibration a curve obtained relating the needle deflection to force. Calibration is carried out by hanging microweights on the needles and measuring the movement. All operations are done with micromanipulators and observed with a microscope.

With unfertilized *Arbacia* eggs, Norris obtained the curves illustrated in Fig. 7 for the tension of a single egg, plotted as a function of increase in length and also of increase

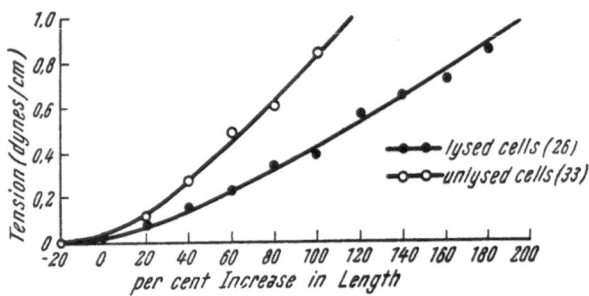

Fig. 8. The average tensions of 33 normal unlysed and 26 mechanically lysed erythrocytes of the salamander, *Triturus pyrrhogaster*, determined by stretching with glass needles. After Norris (1939).

in surface area. The values agree well with those of Cole (1932). Elasticity of the surface is evident from the graph.

A similar procedure, using the oval nucleated erythrocytes of the salamander, *Triturus pyrrhogaster* (dimensions about $43 \times 27 \mu$) gave the curve illustrated in Fig. 8. It will be observed that the surface is elastic and that the tension of a lysed corpuscle, either by puncture or by treatment with saponin is less than that of the intact cell.

## Sessile drop method

The first attempt to measure surface forces of cells by this method was made by VLÈS (1926) who observed unfertilized sea urchin eggs (*Paracentrotus lividus*) from the side and noted an eccentricity which he attributed to the distorting force of gravity resisted by the tension at the surface of the egg. His equation for calculating the surface energy *e* is:

$$e = 10 \sqrt{\frac{b}{a-b}(d_e - d_m)}$$

where *a* is the horizontal and *b* the vertical diameter of the flattened egg, $d_e$ is the density of the egg and $d_m$ the density of sea water. VLÈS obtained values of 15–25 dynes per cm for the tension, but this result appears to be due to the use of an incorrect equation or to previous flattening due to compression in the ovary, the eggs merely settling with their flattened surfaces parallel to the plane on which they rest.

The various equations relating the form of a flattened drop resting on a plane to the tension at its surface have been summarized and discussed by DORSEY (1928, 1929). In general the relation

$$T = g\,(\varrho - \varrho')\, r^2\, F$$

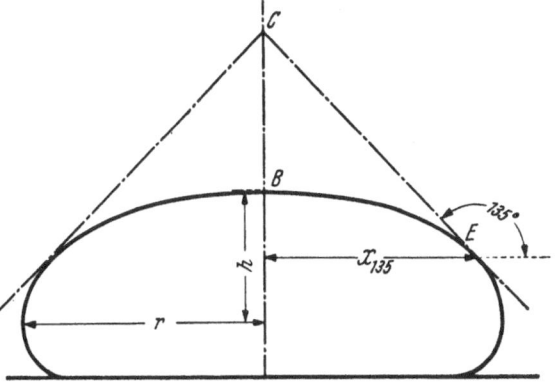

Fig. 9. Diagram of a flattened drop, showing measurements to be made to determine tension by sessile drop method. After DORSEY (1928).

holds. *T* is the tension, *g* the force of gravity, $(\varrho - \varrho')$ the difference in density between drop and medium, *r* the radius of greatest flattening and *F* a function containing *f*, a term representing the flattening of the drop. SHAPIRO (HARVEY and SHAPIRO 1934) has given a table relating *F* and *f* for different degrees of flattening, obtained from certain measurements of the drop profile shown in Fig. 9. These equations strictly apply only where pure surface tension is involved, *i. e.* only where the tension is independent of the extension of the surface. It is easy to calculate that most marine eggs are far too small to flatten appreciably under the influence of gravity even though their surface tension is assumed to be 0.2 dyne/cm (HARVEY 1933).

The large egg of the mollusc, *Busycon canaliculatum*, 1 mm in diameter, does flatten under the influence of gravity and gives a tension of 0.5 dyne/cm, while the egg of the salamander, *Triturus viridescens*, 1.5 mm in diameter, gives 0.1 dyne/cm (HARVEY and FANKHAUSER 1933). The protecting membranes were of course first removed. Both eggs cleave completely so that any membrane at their surface must be non-rigid if not completely liquid in character. It is obvious that sessile drop equations cannot be applied if the surface membrane possesses rigidity (see HARVEY 1936).

## The Interfacial Tension Oil-Protoplasm

The most important application of the sessile drop equations is to small spherical oil drops in living cells, made possible by the development of

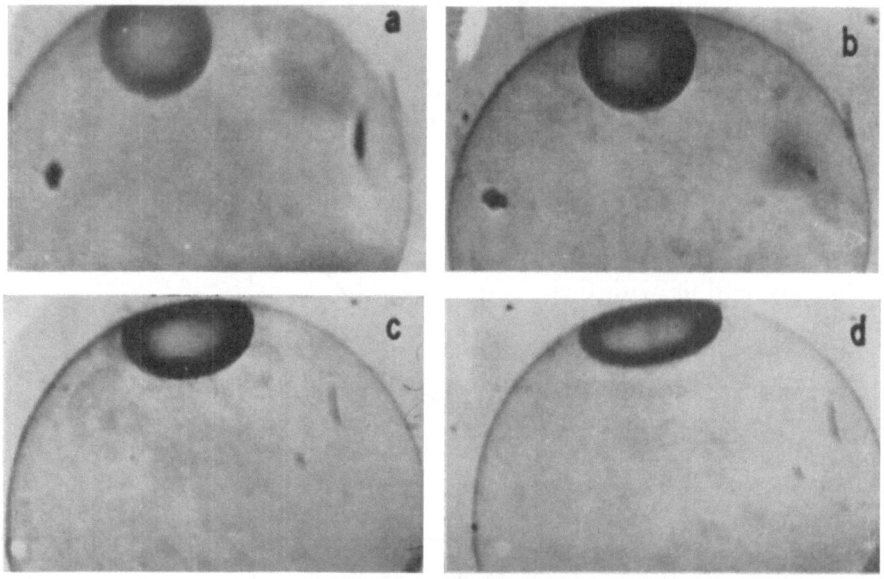

Fig. 10. A natural drop of oil in the egg of the mackerel, photographed through the centrifuge microscope at 7 (a), 62 (b), 195 (c) and 400 (d) times gravity. Note the progressive flattening as the centrifugal force is increased.
After HARVEY and SHAPIRO (1934).

the centrifuge-microscope (HARVEY 1932). Mackerel eggs contain a single droplet, 310 $\mu$ in diameter, which flattens against the rigid egg membrane if the eggs are centrifuged. If photographed while revolving beautiful sharp profiles of flattened drops are obtained (Fig. 10), from which HARVEY and SHAPIRO (1934) calculated the oil-protoplasm interfacial tension to average 0.6 dyne/cm. When the force was increased from 50 to 450 times gravity little change in the tension occurred, showing that the surface did not possess *marked* elastic properties. Measurements of mackerel oil against water gave values of 10 dynes/cm and a study of the oil-water tension at various pH values showed that the low value could not be explained by the pH of the egg. Some strongly surface active substance was indicated, which turned out to be protein (DANIELLI and HARVEY 1935).

On the other hand a study of natural oil drops in the eggs of *Daphnia pulex* (Fig. 11) by Harvey and Schoepfle (1939), indicated that the interfacial tension becomes greater as the area of the drop increases, due to greater flattening with higher centrifugal forces. The effect is shown in Fig. 12. It will be noticed that the interfacial tension ranges from 0.4 to 1.4 dynes/cm.

Measurements of flattened drops of olive oil injected into *Amoeba pro-*

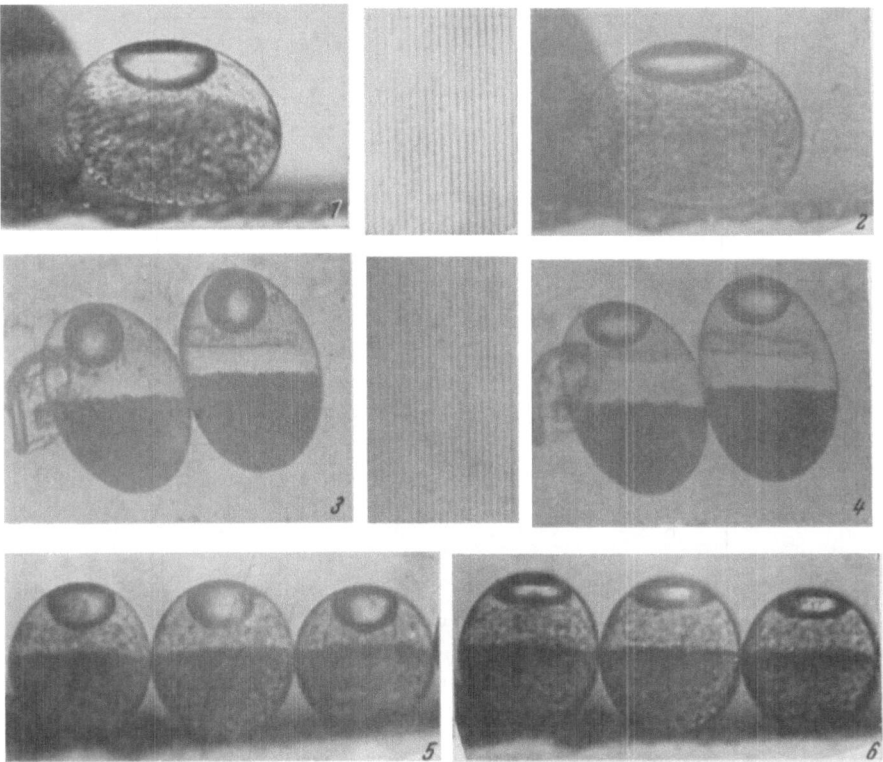

Fig. 11. *Daphnia pulex* eggs containing a single oil drop centrifuged at 215000 × gravity to stratify the granules and then centrifuged at lower forces in the centrifuge microscope to study flattening of the oil drop. The scale between photographs has lines 0.01 mm apart. *1*—1010 × g. *2*—Same egg, 2690 × g. *3*—Not revolving. *4*—Same eggs, 650 × g. *5*—460 × g. *6*—Same eggs, 4690 × g.
After Harvey and Schoepfle (1939).

*teus* and flattened by centrifugal force against the gelled ectoplasmic layer give interfacial tension of 1.8 to 2 dynes/cm.

Kopac's (1940, 1943, 1950) studies on oil droplets and cells has led to a new method of measuring the interfacial tension of oil and protoplasm and the properties of cytoplasmic proteins at the oil surface. The procedure, called the flow-pressure method, is a modification of the maximum bubble pressure method often used for gas-liquid surface tension determination. Oil is forced from a micropipette into the interior of a cell and the pressure $P$ noted as the oil flowing out of the tip increases in size. The pressure will be greatest when the curvature of the oil surface in contact with cytoplasm is greatest (*i. e.* when the radius is least), and the radius is least when the oil is a hemisphere at the end of the tube.  Under these conditions:

$$P = 2\,\gamma/r$$

As more oil moves out of the pipette, the radius of the drop increases and the pressure falls.  Kopac found the same low values for the oil-cytoplasm interface within *Arbacia* eggs that have been observed with

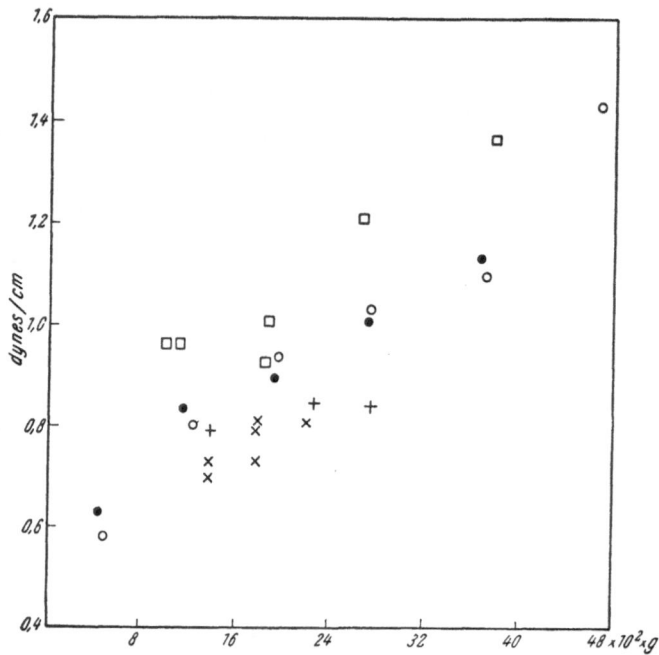

Fig. 12.  Interfacial tension (by sessile drop method) of oil protoplasm in *Daphnia pulex* eggs, as a function of centrifugal force.  Symbols are for different eggs.  The dots are for one egg as the force is increased and circles for the same egg as the force is decreased.
After Harvey and Schoepfle (1939).

mackerel egg oil and *Daphnia* egg oil in contact with cytoplasm, fractions of a dyne/cm.

Highly interesting conclusions can be drawn from the studies of Chambers and Kopac (1937) and Kopac and Chambers (1937) on spontaneous coalescence of oils and sea urchin eggs.  Under certain conditions the egg and oil drop can be treated as two drops of oil uniting on contact.  The oil must be free of adsorbed protein and the surface of the eggs washed in NaCl or KCl to remove any extraneous films [5].  Then if a minute drop of oil from a micropipette is touched to the surface and the volume increased to a critical size, the oil drop penetrates the cell spontaneously, provided the interfacial tension against sea water is greater than (approximately) 9.5 dynes/cm.  Thus, apolar oils penetrate at all *p*H values if no protein is

---

[5] If the surface is not thoroughly washed or is gelled, "capping" rather than penetration occurs (Dawson and Belkin 1929, Marsland 1933, Chambers 1935).

adsorbed on them and polar oils penetrate in more acid $p$H values. In calcium salt solutions the penetration is more rapid.

When an oil drop penetrates, the change is from system $A$ (cell in water and oil in water) to system $B$ (oil in cell in water). If the free energy of system $B$ is less than that of $A$, a spontaneous change of $A$ to $B$ should occur (TAYLOR 1921) *provided the surface is liquid*. From the area of the spheres, measured values for the interfacial tension of oil/sea water and the previously mentioned values for *Arbacia* egg/sea water and oil/protoplasm interfacial tensions, it can be shown that a decrease in free energy does actually occur.

## Discussion and Conclusions

Since it is obvious that something insoluble in water is present at a cell surface, QUINCKE's idea of a lipid surface layer has received support from various directions. OVERTON's lipoid permeability hypothesis has already been mentioned. In explanation of the antagonistic action of salts on cells, particularly sodium and calcium, CLOWES (1916) suggested that the cell might be visualized as an emulsion of oil and other particles in water making up the cell interior, with a water in oil emulsion at the surface. Oil would then be presented as the continuous phase to the external medium. On cytolysis, the water in oil at the surface might change to an oil in water emulsion, with consequent dissolution of the cell. Such reversions can be demonstrated experimentally in oil-water emulsions, brought about by changing the ratio of sodium and calcium in solution. CLOWES pointed out that Na-Ca ratios, which profoundly affect the vital activity of cells and often destroy them, were also critical ratios for reversal of an artificial oil-water emulsion. An ingenious diagram made the surface structure easy to visualize and added plausibility to the hypothesis.

Measurements of electrical capacity made during cell impedance studies (FRICKE 1925) indicate a dielectric oil-like layer of molecular thickness at the cell surface, and the remarkable behavior of oil drops, previously described, when brought in contact with a cell likewise point to abundant surface lipid. Finally the fact that the tension at the surface of cells and the interfacial tension of oil-protoplasm have practically the same low values again suggests that lipid must be an important element in surface structure.

In an analysis of the significance of the low oil-protoplasm interfacial tensions, DANIELLI and HARVEY (1935) studied the interfacial tension of egg oil extracted from mackerel eggs against various aqueous solutions, including egg extract, by means of the DU NOUY tensimeter. It was found that low tensions (0.8 dyne/cm) were obtained with egg oil in contact with egg extracts, but much higher ones with egg oil in contact with sea water (7 dynes/cm). Further analysis of the cause of the low tension pointed to a globulin-like protein as the surface active substance. The oil-drop surface inside the mackerel egg can be pictured as made up of oriented oil molecules, the polar groups towards the water phase, covered with an adsorbed monolayer of hydrated protein molecules.

Since the tension at the surface of cells is as low as that of oil-protein solutions, the inference is obvious that the plasma membrane may be constituted in a similar manner, a thin oil layer, perhaps two or three molecules thick, with protein on each side. Danielli and Davson's (1935) diagram of such a structure is reproduced in Fig. 13, which merely represents a doubling for a thin film of oil of the adsorption of undenatured protein molecules on the polar groups of a fatty acid. Danielli and Davson (1935) and Danielli (1936) have found such films to have considerable stability. Further evidence for and discussion of this "paucimolecular oil film" theory will be found in the reviews of Harvey and Danielli (1938), Danielli (1938), Danielli and Davson (1942) and Danielli (1951).

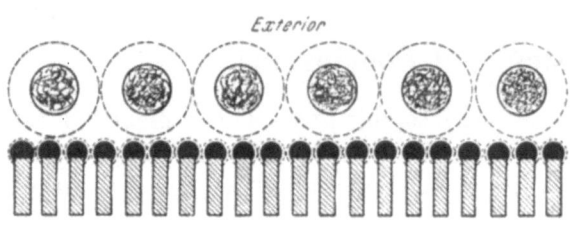

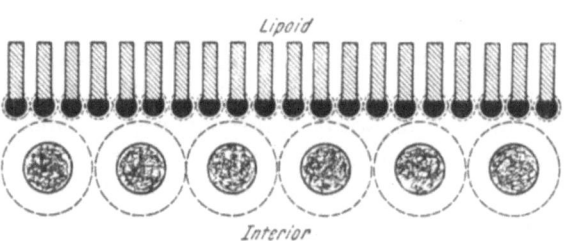

Fig. 13. A possible structure for the plasma membrane of cells, based, among other facts, on the low tensions found at the cell surface, similar to the interfacial tensions of oil-protoplasm. An oriented paucimolecular lipoid layer has adsorbed undenatured hydrated protein molecules on each side.
After Danielli and Davson (1935).

That oil injected into *Arbacia* eggs becomes covered with a monolayer of protein molecules, is indicated if the oil is sucked back into pipette. When the area of the globule of oil decreases to a certain critical size, a wrinkling of the surface occurs (Devaux effect) and the protein film is denatured. This drop retraction technique of Kopac (1940, 1943, 1950) has been used to study cytoplasmic proteins *in vivo* and also *in vitro* after extraction from the cell by various procedures. A great difference in behavior of cell proteins after removal from the cell has been noted, leading to the conclusion that within the cell there are present substances which inhibit denaturation.

There are also other substances which enhance denaturation. These compounds may be formed within the cell under certain conditions. An oil drop injected into an *Asterias* oöcyte does not exhibit a Devaux effect (a crinkling) unless its volume is reduced, but such a droplet will crinkle spontaneously if the egg is cytolysed, or if a local area of cytolysis appears near the oil drop. Such a spontaneous Devaux effect is thought to be due to dissociation of nucleoproteins into nucleic acid and smaller protein molecules as an accompaniment of the cytolytic process. Space does not permit a detailed discussion of the complicated effects but their importance in connection with cell surfaces is obvious. They indicate how tension at that surface may change, provided there is sufficient lipid material present to make its properties similar to that of an oil drop.

In contrast to the lipid theory, many workers have held that the cell surface is protein in composition. No final answer can be given at the present time, but knowledge of surface structure must in final analysis be solved at the molecular level. If one considers the plasma membrane to be made up of a complex mixture of protein and lipid molecules, a "mosaic", PARPART and BALLENTINE (1952) have suggested a possible arrangement for the erythrocyte. Further speculation on the molecular configuration of the plasma membrane is beyond the scope of this paper, except insofar as studies on tension at the surface of cells point toward one or another view.

One question that may be asked is how low can an interfacial tension be? There is apparently no lower limit. DE RUITER and BUNGENBERG DE JONG (1947) have observed interfacial tensions of gum arabic-gelatine coacervate drops ranging from 0.00039 to .0025 dynes/cm and the usual interpretation of crinkling at an oil-water interface is that the tension has fallen to zero (LANGMUIR and WAUGH 1938).

It must be emphasized again that the surface force values for "naked" cells, low as they are, really represent the sum of surface and elastic tensions of a layer made up of a semiliquid "pellicle" and a thin gelled region. The actual surface tension at the boundary of this combination and the external medium may be considerably less. Ever since PLATEAU (1873) showed that a cylinder becomes unstable when its length is greater than $\pi$ times its diameter, then breaking into drops, observers have been disturbed to find long stable pseudopods or filaments projecting from living cells. A solid or gelled core would of course maintain stability, but many of these pseudopodia are definitely liquid, although viscous. The remarkable length of the myelin forms of lecithin is another well known case of stable liquid rods. The explanation of these forms must be sought in very low surface tension and relatively high viscosity. PLATEAU had in mind a liquid with high surface tension and relatively low viscosity, like water. If the viscosity is higher, a cylinder will be stable when its length is many times its diameter, a relation studied by Lord RAYLEIGH (1892).

We are so accustomed to thinking in terms of air-water surface tensions, 73 dynes/cm, that it is hard to realize that the tension of cells is at least 1000, perhaps in many cases 10,000 times less. Where the ratio of surface to volume becomes large, as in small cells, the surface energy (surface tension $\times$ area) becomes a quantity to reckon with. If cells are to grow and divide, work must be done against surface energy. It would be inconvenient for a cell to possess an internal pressure of any magnitude due to surface tension. Cells would not easily deform or adjust their shape to other neighboring cells. A high surface tension might be as disadvantageous to a cell as is a large size, where the slowness of diffusion would limit activity. Since large surface areas are necessary for vital functions, it is not surprising to find that the tension of these surfaces and hence the surface energy has been reduced as much as possible. Although measurements have been made on relatively few cells, all those which have been studied, when tested by the most diverse methods, give tensions which are extraordinarily low. When protection is necessary, a second rigid membrane is developed. The reduction of protoplasmic surface forces to a minimum is merely another example in the molecular realm of adjustment to physical reality observed so frequently in the structure of living organisms.

## Literature

Early historical references are not included unless the year is enclosed in brackets in the text.

Bernstein, J., 1900: Chemotropische Bewegung eines Quecksilbertropfens. Pflügers Arch. **80**, 628—637.

Berthold, G. D. W., 1886: Studien über Protoplasmamechanik. Leipzig.

Bütschli, O., 1876: Studien über die ersten Entwicklungsvorgänge der Eizelle, die Zellteilungsvorgänge der Eizelle, die Zellteilung und die Konjugation der Infusorien. Abh. Senckenberg. naturf. Ges., Frankf. M. **10**, 232—452.

— 1889: Versuche zur Nachahmung von Protoplasmastructuren. Dtsch. Natf. Tagebl. 1889, 266—267; also Quart. J. microsc. Sci. **31**, 99—103, 1890, a letter to E. R. Lancaster.

— 1894: Investigations on Microscopic Foams and on Protoplasm. London, 379 pp., translated by E. A. Minchin from the German ed. 1892.

— 1900: Bemerkungen über Plasmaströmungen bei der Zellteilung. Arch. Entw. **10**, 52—57.

Chambers, R., 1935: Studies on the physical properties of the plasma membrane. Biol. Bull. **69**, 331 (abstract).

— 1940: The relation of extraneous coats to the organization and permeability of cellular membranes. Cold Spring Harbor Symposia **8**, 144—153.

— 1944: Some physical properties of protoplasm. In J. Alexander's Colloid Chemistry, New York. Vol. 5, pp. 864—875.

— and M. J. Kopac, 1937: The Coalescence of living cells with oil drops. I. *Arbacia* eggs immersed in sea water. J. cellul. a. comp. Physiol. (Am.) **9**, 331—344.

Clowes, G. H. A., 1916: Protoplasmic equilibrium. J. Phys. Chem. **20**, 407—451.

Cole, K. S., 1932: Surface forces of the *Arbacia* egg. J. cellul. a. comp. Physiol. (Am.) **1**, 1—9.

— and E. Michaelis, 1932: Surface forces of fertilized *Arbacia* eggs. J. cellul. a. comp. Physiol. (Am.) **2**, 121—126.

Costello, D. P., 1935: Fertilization membranes of strongly centrifuged *Asterias* eggs. Physiol. Zool. **8**, 65—72.

Czapek, F., 1911: Methode der direkten Bestimmung der Oberflächenspannung der Plasmahaut der Pflanzenzellen. Jena; also in Ber. dtsch. Bot. Ges. **28**, 159—169 and 480—487, 1910.

Danielli, J. F., 1936: Some properties of lipoid films in relation to the structure of the plasma membrane. J. cellul. a. comp. Physiol. (Am.) **7**, 393—407.

— 1942, 1952: The cell surface and cell physiology, in: Cytology and Cell Physiology. Ed. by G. H. Bourne, Oxford.

— and H. Davson, 1935: A contribution to the theory of permeability of thin films. J. cellul. a. comp. Physiol. (Am.) **5**, 495—508.

— — 1942: The Permeability of Natural Membranes. Cambridge.

— and E. N. Harvey, 1935: The tension at the surface of mackerel egg oil, with remarks on the nature of the cell surface. J. cellul. a. comp. Physiol. (Am.) **5**, 483—494.

Dawson, J. A., and M. Belkin, 1929: The digestion of oils by *Amoeba proteus*. Biol. Bull. **56**, 80—86.

De Bruyn, P. P. H., 1947: Theories of Amoeboid movement. Quart. Rev. Biol. (Am.) **22**, 1—24.

DE RUITER, L., and H. G. BUNGENBERG DE JONG, 1947: The interfacial tension of gum arabic-gelatine complex coacervates and their equilibrium liquids. Proc. K. Nederl. Akad. van Wetensch. Amsterdam **50**, 836—848.

DE VRIES, G. A., 1947: The influence of lithium chloride and calcium chloride on viscosity and tension at the surface of uncleaved eggs of *Limnaea stagnalis*, 2. Proc. K. Nederl. Akad. van Wetensch. Amsterdam **50** (10), 1335—1442.

DORSEY, W. E., 1928: A new equation for the determination of surface tension from the form of a sessile drop or bubble. J. Wash. Acad. Sci. **18**, 505—509.

— 1929: The investigation of surface tension and associated phenomena. Bull. Nat. Res. Coun., No. 69, 56—118.

ENGLEMANN, W., 1869: Beiträge zur Physiologie des Protoplasmas. Pflügers Arch. **2**, 307—322.

EWART. A. J., 1903: On the Physics and Physiology of Protoplasmic Streaming in Plants. Oxford.

FRICKE, H., 1925: The electric capacity of suspensions with special reference to blood. J. gen. Physiol. (Am.) **9**, 137—152.

FÜRTH, O., 1922: Zur Theorie der amoeboiden Bewegungen. Arch. néerl. Physiol. **7**, 39—43.

GAD, J., 1878: Zur Lehre der Fettresorption. Arch. Anat. usw. 1878 (part 2), 181—205.

GRUBER, K., 1912: Biologische und experimentelle Untersuchungen an *Amoeba proteus*. Arch. Protistenk. **25**, 316—376.

HABERLANDT, L., 1919: Über amoeboide Bewegung. Z. Biol. **69**, 409—436.

HARVEY, E. B., 1932: The development of half and quarter eggs of *Arbacia punctulata* and of strongly centrifuged whole eggs. Biol. Bull. **62**, 155—167.

— 1933: Effects of centrifugal force on fertilized eggs of *Arbacia punctulata* as observed with the centrifuge-microscope. Biol. Bull. **65**, 389—396.

— 1935: Some surface phenomena in the fertilized sea urchin egg as influenced by centrifugal force. Biol. Bull. **69**, 298—304.

— 1945: Stratification and breaking of the *Arbacia punctulata* egg when centrifuged in single salt solutions. Biol. Bull. **89**, 72—75.

HARVEY, E. N., 1931 a: A determination of the tension at the surface of eggs of the anelid, *Chaetopterus*. Biol. Bull. **60**, 67—71.

— 1931 b: The tension at the surface of marine eggs, especially those of the sea urchin, *Arbacia*. Biol. Bull. **61**, 273—279.

— 1932: The Microscope-centrifuge and some of its applications. J. Franklin Inst. **214**, 1—23.

— 1933: The flattening of marine eggs under the influence of gravity. J. cellul. a. comp. Physiol. (Am.) **4**, 35—47.

— 1936: The properties of elastic membranes with special reference to the cell surface. J. cellul. a. comp. Physiol. (Am.) **8**, 251—260.

— 1938: Some physical properties of protoplasm. J. App. Physics **9**, 68—80. See also Trans. Farad. Soc. **33**, 943—946, 1937.

— and J. F. DANIELLI, 1936: The elasticity of thin films in relation to the Cell Surface. J. cellul. a. comp. Physiol. (Am.) **8**, 31—36.

— — 1938: Properties of the cell surface. Biol. Rev. **13**, 319—341.

— and G. FANKHAUSER, 1933: The tension at the surface of the eggs of the salamander, *Triturus (Diemyctylus) viridescens*. J. cellul. a. comp. Physiol. (Am.) **3**, 463—475.

— and D. A. MARSLAND, 1932: The tension at the surface of *Amoeba dubia*, with direct observations on the movement of cytoplasmic particles at high centrifugal speeds. J. cellul. a. comp. Physiol. (Am.) **2**, 75—97.

— and H. SHAPIRO, 1934: The interfacial tension between oil and protoplasm within the living cells. J. cellul. a. comp. Physiol. (Am.) **5**, 255—267.

— — 1941: The recovery period (relaxation) of marine eggs after deformation. J. cellul. a. comp. Physiol. (Am.) **17**, 135—144.

— and G. SCHOEPFLE, 1939: The interfacial tension of intracellular oil drops in the eggs of *Daphnia pulex* and in *Amoeba proteus*. J. cellul. a. comp. Physiol. (Am.) **13**, 383—389.

Hatschek, E., 1910: Die Filtration von Emulsionen und die Deformation von Emulsionsteilchen unter Druck. Kolloid-Z. **7**, 81—86.

Heilbrunn, L. V., 1913—1925: Studies in artificial parthenogenesis. Biol. Bull. **24**, 343—361, 1913; **29**, 149—203, 1915; **46**, 277—280, 1924; **49**, 241—249, 1925.

— 1952: An Outline of General Physiology. 3rd ed. Philadelphia, pp. 75—88.

Herčik, F., 1934: Oberflächenspannung; in: Medicinische Kolloidlehre, vol. I, Dresden.

Hirschfeld, L., 1909: Ein Versuch, einige Lebenserscheinungen der Amoeben physikalisch-chemisch zu erklären. Z. allg. Physiol. **9**, 529—534.

Hoeber, R., 1902—1926: Physikalische Chemie der Zelle und Gewebe. 1st ed. 1902; 2nd ed. 1906; 3rd ed. 1911; 4th ed. 1914; 5th ed. 1922—1924; 6th ed. 1926.

— D. I. Hitchcock, J. B. Bateman, D. R. Goddard and W. O. Fenn, 1945: Physical Chemistry of Cells and Tissues. Philadelphia and Toronto.

Hofmeister, W. F. B., 1867: Die Lehre von der Pflanzen-Zelle. Leipzig. 1st vol. of Handbuch der Physiologischen Botanik, ed. by W. Hofmeister.

Jennings, H. S., 1906: Behavior of the Lower Organisms. New York. 2nd ed. 1923.

Jensen, P., 1901: Untersuchungen über Protoplasmamechanik. Pflügers Arch. **87**, 361—417.

— 1902: Die Protoplasmabewegung. Erg. Physiol. **1** (part 2), 1—42.

Kao, C., R. Chambers and E. L. Chambers, 1951: The internal hydrostatic pressure of the unfertilized *Fundulus* egg activated by puncture. Biol. Bull. **101**, 210—211. See also pp. 206—207.

Kopac, M. J., 1940: The physical properties of the extraneous coats of living cells. Cold Spring Harbor Symposia **8**, 154—170.

— 1943: Micrurgical applications of surface chemistry to the study of the living cell; in: Micrurgical and Germ-free Methods. Springfield, Ill., and Baltimore, Md.

— 1944: Some surface-chemical properties of protoplasmic proteins. In J. Alexander's Colloid Chemistry, New York. Vol. 5, pp. 875—883.

— 1950: The surface chemical properties of cytoplasmic proteins. Ann. N. Y. Ac. Sc. **50**, 870—909.

— and R. Chambers, 1937: Coalescence of living cells with oil drops. II. *Arbacia* eggs immersed in acid or alkaline calcium solution. J. cellul. a. comp. Physiol. (Am.) **9**, 345—362.

Kopaczewski, W., 1933: Rôle de la tension superficielle en biologie. Protoplasma **19**, 255—292.

Krizenecky, J., and O. Dubiska, 1927: Eine Methode zur Messung der Oberflächenspannung biologischer Flüssigkeiten gegen ein Protoplasma-ähnliches Medium. Protoplasma **2**, 460—497.

Langmuir, I., and D. Waugh, 1938: The adsorption of proteins at oil-water interfaces and artificial protein-lipoid membranes. J. gen. Physiol. (Am.) **21**, 745—755.

Lillie, R. S., 1903: Fusion of blastomeres and nuclear division without cell division in solutions of non-electrolytes. Biol. Bull. **4**, 164—178.

— 1909: The general biological significance of changes in permeability of the surface layer or plasma membrane of living cells. Biol. Bull. **17**, 188—208.

Marsland, D. A., 1933: The site of narcosis in a cell: the action of a series of paraffin oils on *Amoeba dubia*. J. cellul. a. comp. Physiol. (Am.) **4**, 9—34.

— 1942: Protoplasmic streaming in relation to gel structure in the cytoplasm. In The Structure of Protoplasm, pp. 127—161. Ames, Iowa.

— 1951: The action of hydrostatic pressure on cell division. Ann. N. Y. Ac. Sc. **51**, 1327—1335.

McClendon, J. F., 1910: On the dynamics of cell division. Amer. J. Physiol. **27**, 240—275.

-- 1911: Ein Versuch, amoeboide Bewegung als Folgeerscheinung des wechselnden elektrischen Polarisationszustandes der Plasmahaut zu erklären. Pflügers Arch. **140**, 271—280.

— 1912: A note on the dynamics of cell division. Arch. Entw. **34**, 263—266.

Mitchison, J. M., 1952: Cell membranes and cell division. Symposia Soc. Exp. Biol. **6**, 105—127.

Moore, A. R., 1930: Fertilization and development without membrane formation in the eggs of the sea urchin, *Strongylocentrotus purpuratus*. Protoplasma **9**, 9—17.

Norris, C. H., 1939: The tension at the surface and other physical properties of the nucleated erythrocyte. J. cellul. a. comp. Physiol. (Am.) **14**, 117—133.

Parpart, A. K., and R. Ballantine, 1952: Molecular anatomy of the red cell membrane; in: Trends in Physiology and Biochemistry. Ed. by E. S. G. Barron, New York, pp. 135—148.

Pfeffer, W., 1891: Zur Kenntnis der Plasmahaut und der Vacuolen etc. Abh. d. math.-phys. Kl. d. Sächs. Akad. d. Wiss. **16**, 185—344. See p. 264 and 266.

Pfeiffer, H., 1935: Über die mechanische Deformierung nackter Protoplasma-Blasen. Protoplasma **23**, 210—216.

— 1936 a: Beiträge zur quantitativen Bestimmung von Molekularkräften des Protoplasmas. IV. Eine Methode zur Bestimmung der Oberflächenspannung nackter Protoplasten gegen ein flüssiges Medium. Protoplasma **25**, 397—403; see also 528—545.

— 1936 b: Further tests of the elasticity of protoplasm. Physics **7**, 302—305.

— 1936 c: Beiträge zur quantitativen Bestimmung von Molekularkräften des Protoplasmas. V. Versuche zur Ermittlung eines Youngeschen „Dehnungs-moduls" von Protoplasmablasen. Protoplasma **26**, 327—376.

— 1937: Experimental researches on the non-Newtonian nature of protoplasm. Internat. J. Cytology. Fujii Jubilee, Vol. 701—710.

Quincke, G. H., 1870: Über Capillaritätserscheinungen an der gemeinschaftlichen Oberfläche zweier Flüssigkeiten. Pogg. Ann. d. Phys. **139**, 1—89; also Phil. Mag. (4th ser.) **41**, 245—266; 370—390; 354—476.

— 1877: Über den Randwinkel und die Ausbreitung von Flüssigkeiten auf festen Körpern. Wied. Ann. Phys., N. F. **2**, 145—194; also Phil. Mag. (5th ser.) **5**, 321—339; 415—433, 1878.

— 1888: Über periodische Ausbreitung an Flüssigkeits-Oberflächen und dadurch hervorgerufene Bewegungserscheinungen. Wied. Ann. Phys., N. F. **35**, 580—642.

Raven, C. P., 1945: The development of the egg of *Limnaea stagnalis* L. from oviposition till first cleavage. Arch. néerl. Zool. **7**, 91—121.

Rayleigh, L., 1892: On the stability of a cylinder of viscous liquid under capillary force. Phil. Mag. (5th ser.) **34**, 145—154.

Rhumbler, L., 1896: Versuch einer mechanischen Erklärung der indirekten Zell- und Kerntheilung. Arch. Entw. **3**, 526—623.

— 1898: Physikalische Analysis von Lebenserscheinungen der Zelle. I. Bewegung etc. Arch. Entw. **7**, 103—198 and Erg. Anat. Entw. **8**, 543—625.

— 1899: Physikalische Analyse von Lebenserscheinungen der Zelle. Arch. Entw. **9**, 33—102.

— 1905: Zur Theorie der Oberflächen-Kräfte der Amöben. Z. wiss. Zool. **83**, 1—52.

— 1914: Das Protoplasma als physikalisches System. Erg. Physiol. **14**, 474—617.

Robertson, T. B., 1909: Note on the chemical mechanics of cell division. Arch. Entw. **27**, 29—34.

— 1911: Further remarks on the chemical mechanics of cell division. Arch. Entw. **32**, 308—313.

Runnström, J., 1952: The cell surface in relation to fertilization. Symposia Soc. Exp. Biol. **6**, 39—88.

— and L. Monné, 1945: On some properties of the surface-layers of immature and mature sea-urchin eggs, especially the changes accompanying nuclear and cytoplasmic maturation. Arkif. Zool. **36 A** (No. 18), 26 pp.

— — and E. Wickland, 1946: Studies on the surface layers and the formation of the fertilization membrane in sea urchin eggs. J. Colloid Sci. **1**, 421—452.

Shapiro, H., 1941: Centrifugal elongation of cells, and some conditions governing the return to sphericity, and cleavage time. J. cellul. a. comp. Physiol. (Am.) **18**, 61—78.

— and E. N. Harvey, 1936: The tension at the surface of macrophages. J. cellul. a. comp. Physiol. (Am.) **8**, 21—30.

Sichel, F. J. M., and A. C. Burton, 1936: A kinetic method of studying surface forces in the egg of *Arbacia*. Biol. Bull. **71**, 397—398.

Spek, J., 1918: Oberflächenspannungsdifferenzen als eine Ursache der Zellteilung. Arch. Entw. **44**, 5—113.

Swann, M. M., 1952: The nucleus in fertilization and cell-division. Symposia Soc. Exp. Biol. **6**, 89—104.

Tait, J., 1920: Capillary phenomena observed in blood cells: Thigmocytes, phago-cytosis, amoeboid movement, differential adhesiveness of corpuscles, emigration of leucocytes. Quart. J. exper. Physiol. **12**, 1—33.

Taylor, W., 1921: The coalescence of liquid spheres–molecular diameters. Phil. Mag. **41**, 877—889.

Thompson, D. W., 1942: Growth and Form. 2nd ed. Cambridge. 1st ed. 1916.

Tiegs, O. W., 1928: Surface tension and the theory of protoplasmic movement. Protoplasma **4**, 88—139.

Verworn, M., 1892: Die Bewegung der lebendigen Substanz. Jena.

— 1895: Vergleichende Physiologie. Jena. Trans. by F. S. Lee. London and New York 1899.

Vexler, D., 1935: A value for the tension at the surface of a myxomycete. Proc. Soc. exper. Biol. a. Med. (Am.) **32**, 1539—1541.

Vlès, F., 1926: Les tensions de surface et les déformations de l'œuf d'Oursin. Arch. Physique biol. **4**, 263—284.

— 1933: Recherches sur une déformation mécanique des œufs d'Oursin. Arch. Zool. exp. et gén. **75**, 421—463.

Wilson, E. B., 1925: The Cell in Development and Heredity. New York. 3rd ed., pp. 192—197.

Wilson, W. L., 1951: The rigidity of the cell cortex during cell division. J. cellul. a. comp. Physiol. (Am.) **38**, 409—416.

— and L. V. Heilbrunn, 1952: The protoplasmic cortex in relation to stimulation. Biol. Bull. **103**, 139—144.

Wright, T. S., 1867: Observations on British zoophytes and protozoa. J. Anat. a. Physiol. **1**, 332—338.

*Soeben hat zu erscheinen begonnen:*

# Protoplasmatologia
## Handbuch der Protoplasmaforschung

Unter Mitwirkung von

**W. H. Arisz**, Groningen · **H. Bauer**, Wilhelmshaven · **J. Brachet**, Bruxelles · **H. G. Callan**, St. Andrews
**R. Collander**, Helsinki · **K. Dan**, Tokyo · **E. Fauré-Fremiet**, Paris · **A. Frey-Wyssling**, Zürich · **L. Geitler**,
Wien · **K. Höfler**, Wien · **M. H. Jacobs**, Philadelphia · **D. Mazia**, Berkeley · **A. Monroy**, Palermo
**J. Runnström**, Stockholm · **W. J. Schmidt**, Gießen · **S. Strugger**, Münster

Herausgegeben von

Prof. Dr. **L. V. Heilbrunn** und Prof. Dr. **F. Weber**
Philadelphia                            Graz

In 14 Bänden

*Das Handbuch erscheint in selbständigen Einzelveröffentlichungen, die*
*in kurzen Zeitabständen aufeinanderfolgen und zu Bänden vereinigt*
*werden. Jeder selbständig erscheinende Handbuchteil ist einzeln käuflich.*
*Über die Disposition des Gesamtwerkes und die nächsten Veröffentlichungen*
*gibt der Verlag bereitwilligst Auskunft.*

*Bei Verpflichtung zur Abnahme des gesamten Handbuches, bei*
*Vorbestellung der einzelnen Teile sowie für Abonnenten der*
*Zeitschrift „Protoplasma" ermäßigt sich der Preis um 20%*

*Bisher sind erschienen:*

**Endomitose und endomitotische Polyploidisierung.** Von Prof. Dr. **Lothar Geitler**,
Botanisches Institut der Universität Wien. **Band VI. Kern- und Zellteilung**, C. Endomitose und endo-
mitotische Polyploidisierung. Mit 44 Textabbildungen. IV, 89 Seiten. 1953.
S 140.—, DM 23.50, $ 5.60, sfr. 24.10

**Chemistry and Physiology of Mitochondria and Microsomes.** By Olov Lindberg,
Ph. D., and **Lars Ernster**, Ph. D., beide Wenner-Gren's Institute, Stockholm. **Band III. Cytoplasma-
Organellen**, A. Chondriosomen, Mikrosomen, Sphaerosomen, 4. Chemistry and Physiology of Mitochondria
and Microsomes. With 32 Figures. IV, 136 Pages. 1954.          S 204.—, DM 34.—, $ 8.10, sfr. 34.80

*Anschließend werden erscheinen:*

**Makromolekulare Chemie und ihre Bedeutung für die Protoplasmaforschung.**
Von Nobelpreisträger Prof. Dr. phil., Dr.-Ing. e. h., Dr. rer. nat. h. c. **Hermann Staudinger** und Dr. phil.,
Mag. rer. nat. **Magda Staudinger**, beide Staatliches Forschungsinstitut für makromolekulare Chemie der
Universität Freiburg i. Br. **Band I. Grundlagen**, 1. Makromolekulare Chemie und ihre Bedeutung für die
Protoplasmaforschung. Mit 27 Textabbildungen. Etwa 80 Seiten. 1954.
**Vorbestellpreis, gültig bis zum Erscheinen:** S 93.60, DM 15.60, $ 3.70, sfr. 15.90
Endgültiger Ladenpreis nach Erscheinen: S 117.—, DM 19.50, $ 4.65, sfr. 20.—

**The Enzymology of the Cell Surface.** By Aser Rothstein, Division of Pharmacology, Depart-
ment of Radiation Biology, University of Rochester School of Medicine and Dentistry, Rochester, New York.
With 21 Figures. Pages I—IV, 1—86. — **Tension at the Cell Surface.** By E. Newton Harvey,
Biology Department, Princeton University, Princeton, New Jersey, USA. With 13 Figures. Pages 1—30. 1954.
(**Band II: Cytoplasma.** E. Cytoplasma-Oberfläche. 4. The Enzymology of the Cell Surface. 5. Tension at
the Cell Surface.)          **Vorbestellpreis, gültig bis zum Erscheinen:** S 134.40, DM 22.40, $ 5.35, sfr. 23.—
Endgültiger Ladenpreis nach Erscheinen: S 168.—, DM 28.—, $ 6.70, sfr. 28.80

*Bei Verpflichtung zur Abnahme des Gesamtwerkes gilt der Vorbestellpreis weiter als Subskriptionspreis.*